Electrical Engineering 101

Electrical Engineering 101

Everything You Should Have Learned in School, but Probably Didn't

by Darren Ashby

AMSTERDAM • BOSTON • HEIDELBERG • LONDON
NEW YORK • OXFORD • PARIS • SAN DIEGO
SAN FRANCISCO • SINGAPORE • SYDNEY • TOKYO

Newnes is an imprint of Elsevier

Newnes is an imprint of Elsevier
30 Corporate Drive, Suite 400, Burlington, MA 01803, USA
Linacre House, Jordan Hill, Oxford OX2 8DP, UK

Recognizing the importance of preserving what has been written, Elsevier prints its books on acid-free paper whenever possible.

Library of Congress Cataloging-in-Publication Data

Application submitted

British Library Cataloguing-in-Publication Data
A catalogue record for this book is available from the British Library.

ISBN-13: 978-0-75067-812-4
ISBN-10: 0-7506-7812-7

For information on all Newnes publications
visit our website at www.books.elsevier.com.

06 07 08 09 10 10 9 8 7 6 5 4 3 2

Printed in the United States of America

I'd like to dedicate this book to my wife Denise;
without her it would have not happened.
She is my passion, my friend and my confidant.

Contents

Preface

The First Word

The intent of this book is to cover the basics that I believe have been either left out of your education or forgotten over time. Hopefully, it will become one of those well-worn texts that you drop on the desk of the new guy when he asks you a question. There is something for every student, engineer, manager, and teacher in electrical engineering. My mantra is, "It ain't all that hard!" Years ago I had a counselor in college tell me proudly that they flunked out over half of the students that started the engineering program. Needing to stay on her good side, I didn't say much at the time. I always wondered, though. If you fail so many students, isn't that really a failure to teach the subject well? I say "It ain't all that hard" to emphasize that even a hick with bad grammar like me can understand the world of electrical engineering. This means you can too! I take a different stance than that counselor of years ago, asserting that everyone who wants to can understand this subject. I believe that many more than 50% of the people who read this book will get something out of it. It would be nice to show the statistics of that to that counselor someday; she was encouraging me to drop out when she made her comment. So, good luck, read on and prove me right—it ain't all that hard!

One more item of note: at the end of each topic I have created some bullet points I like to call 'thumb rules." They are what they seem—those "rule of thumb" concepts that the really good engineers just seem to know. These concepts are what always lead them to the right conclusion and a solution to the problem. If you get bored with a section, make sure to hit the thumb rules anyway. There you will get the distilled core concepts that you absolutely should know. Thank you for giving me a chance to pass on a few ideas. I hope you find them engaging, fun, and educational!

Overview

For Engineers

Granted, there are many good teachers out there and you may have gotten the basics, but time and too many "status reports" have dulled the finish on your basic knowledge set. If you are like me, you have found a few really good books that you often pull off shelf in time's of need. They usually have a well-written, easy-to-understand explanation of the particular topic you need to apply. I hope this will be one of those books for you.

You may also be a fish out of water, an ME thrown into the world of electrical engineering, and you would like a basic understanding to work with the EEs around you. If you get a good understanding of these principles, I guarantee you will surprise at least some of the "sparkies" (as I like to call them) with your intuitive insights to problems at hand.

For Students

I don't mean to knock the collegiate educational system, but it seems to me that too often student's can pass a class in school with the "assimilate and regurgitate" method. You know what I mean—go into class, soak up all the things the teacher wants you to know, take the tests, say the right things at the right time, and leave the entire class without an ounce of applicable knowledge. I think many students are forced into this mode when teachers do not take the time to lay the groundwork for the subject they are covering. Students are so hard-pressed just to keep up that they do not get the light bulb to go on over their head. The reality is, if you leave the class with a fundamental, intuitive understanding of the topic, you will be imminently more successful applying that basic knowledge than anything from the end of the syllabus for that class.

For Managers

The job of the engineering manager should have more to it than what is depicted by the pointy-haired boss you see in the Dilbert® cartoons. One thing many managers do not know about engineers is that they welcome truly insightful takes on whatever they are working on. Please notice I said "truly insightful"; you can't just spout off some acronym you heard in the lunchroom and expect your engineers to pay attention. However, if you understand these basics, I am sure there will be times when you will be able to point your engineers in the right direction. You will be happy to keep the project moving forward, and they will gain a new respect for their boss. (They might even put away their pointy-haired doll!)

For Teachers

Please don't get me wrong—I don't mean to say all teachers are bad. In fact (except for one or two) my teachers were really good instructors. However, sometimes I think the system is flawed. Given pressures from the dean to cover X, Y, and Z topics, sometimes the more fundamental X and Y are sacrificed just to get to topic Z. I did get a chance to teach a semester at my own alma mater, and some of these chapters are directly from that class. My hope for teachers is to give you another tool that you can use to flip the switch on the "Aha!" light bulb over your students' heads.

Acknowledgments

No book is a single person's work. What any individual knows is a culmination of experience and lessons from many sources. I would like to thank some of the individuals that made this text possible.

My father for his common sense and his living proof that you can do what others say is impossible.

My mother for the smart genes she passed on to me and for her undying belief in the perfection of her children.

My wife for her patience and encouragement and doing the things I should have been doing just so I could get another chapter written.

My kids for understanding why dad couldn't make it to every game and had to spend time in his office every night.

My brother Robert with whom I embarked on this literary adventure, and Steve Petersen, a close friend and aspiring engineer for the late night graphic work.

For any of you I missed and feel neglected in not being thanked, give me a call. I'll buy you lunch!

What's on the CD-ROM?

The CD-ROM includes some handy files:

Circuit Simulator Link from Electronics Workbench™

Used properly, a simulator is a great way to check out your knowledge and help you understand how these parts work. Make sure you use it for the basic stuff first and read the chapter on tools!

Schematic Capture and PCB Layout

A handy tool to prototype your design. The company that offered this will let you upload the files right to their site and turn PCBs at reasonable price too!

FilterPro™

Filter pro, a cool program from TI to help you design op-amp filters of all types.

Thumb Rules Reference

A PDF reference of all the thumb rules in the text so that you can print out and stick all over your walls to impress the ladies. OK so it probably won't work, but it will help you remember these secrets of success!

Equations Reference

A PDF referencing all the equations used in the text. Put this on the wall next to the thumb rules and people will think you are pretty smart knowing all that stuff. The real benefit is that you won't have to remember anything and can save your brain power for more important things, like pondering what you should have for lunch.

CHAPTER 1

Three Things They Should Have Taught in Engineering 101

Do you remember your engineering introductory course? At most, I'll venture that you are not sure you even had a 101 course. It's likely that you did and, like the course I had, it really didn't amount to much. In fact, I don't remember anything except that it was supposed to be an "introduction to engineering."

Much later on in my senior year, and shortly after I graduated, I learned some very useful general engineering methodologies. They are so beneficial that I sincerely wish they had taught these three things from the beginning of my course work. In fact, it is my belief that this should be basic, BASIC knowledge that any aspiring engineer should know. I promise that by using these in your day-to-day challenges you will be more successful and, besides that, everyone you work with will think you are a genius. If you are a student reading this, you will be amazed at how many problems you can solve with these skills. They are the fundamental building blocks for what is to come.

Units Count!

This is a skill that one of my favorite teachers drilled into me my senior year. Till I understood this, I forced myself to memorize hundreds of equations just to pass tests. After applying this skill I found that, with just a few equations and a little algebra, you could solve nearly any problem. This was definitely an "Aha" moment for me. Suddenly the world made sense. Remember those dreaded story problems that you had to do in physics? Using unit math, those problems became a breeze; you can do them without even breaking a sweat.

Unit Math

With this process the units that the quantities are in become very important. You don't just toss them aside because you can't put them in your calculator. In fact, you figure out the units you want in your answer and then work the problem backwards to figure out what you need to solve it. You do all this before you do anything with the numbers at all. The basic concept of this was taught way back in algebra, but no one told you to do it with the units. Let's look at a very simple example.

You need to know how fast your car is moving in miles per hour (mph). You know it traveled one mile in one minute. The first thing you need to do is figure out the units of the answer. In this case it is mph or miles per hour. Now write that down (remember "per" means divided by).

$$answer = something \cdot \frac{miles}{hour}$$

Now arrange the data that you have in a format that will give you the units you want in the answer.

$$1 \cdot mile \times \frac{1}{1 \cdot \text{min}} \times \frac{60 \cdot \text{min}}{1 \cdot hour} = answer$$

Remember, whatever is above the dividing line cancels out whatever is the same below the line, something like this:

$$1 \cdot mile \times \frac{1}{1 \cdot \cancel{\text{min}}} \times \frac{60 \cdot \cancel{\text{min}}}{1 \cdot hour} = answer$$

When all of the units that can be removed are gone, what you are left with is 60 mph, which is the correct answer. Now you might be saying to yourself that was easy. You are right! That is the point after all—we want to make it easier. If you follow this basic format, most of the "story problems" you encounter every day will bow effortlessly to your machinations.

Another excellent place to use this technique is for solution verification. If the answer doesn't come out in the right units, most likely something was wrong in your calculation. I always put units on the numbers and equations I use in MathCad® (a tool no engineer should be without.) To see the correct units when it is all said and done confirms that the equations are set up properly. (The nice thing is MathCad automatically handles the conversions that are often needed.) So, whenever you come upon a question that seems to have a whole pile of data and you have no idea where to begin, first figure out what units you want the answer in. Then, shape that pile of data till the units match the units needed for the answer.

Remember this: *by letting the units mean something in the problem, the answer you get will actually mean something too.*

Sometimes Almost Is Good Enough

My father had a saying, "Almost only counts in horseshoes and hand grenades!" He usually said this right after I "almost" put his tools away or I "almost" finished cleaning my room. Early in life I became somewhat of an expert in the field of "almost." As my dad pointed out, there are many times when almost doesn't count. However, as the saying says, it probably is good enough to almost hit your target with a hand grenade. There are a few other times when almost is good enough too. One of them is when you are trying to estimate a result. A skill that goes hand in hand with the idea of unit math is that of estimation.

The skill or art of estimation involves two main points. The first is rounding to an easy number and the second is understanding ratios and percentages.

The rounding part comes easy. Let's say you are adding two numbers, 97 and 97. These are both nearly 100, so say they are 100 for a minute, add them together and you get 200, or nearly so. Now this is a very simplified explanation of this idea, and you may think "Why didn't you just type 97 into your calculator a couple of times and press equal"? The reason is, as the problems become more and more complex, it becomes easier to make a mistake that can cause you to be far off in your analysis. Let's apply this idea to our previous example. If your calculator says 487 after you added 97 to 97, and you compare that with the estimate of 200 that you did in your head, you quickly realize that you must have hit a wrong button.

Ratios and percentages help get an idea of how much one thing affects another. Say you have two systems that add their outputs together. In your design, one system outputs one hundred times more than the other. The ratio of one to the other is 100:1. If the output of this product is way off, which of these two systems do you think is most likely at fault? It becomes obvious that one system has a bigger effect when you estimate the ratio of one to the other.

Developing the skill of estimation will help eliminate hunting dead ends and chasing your tail when it comes to engineering analysis and troubleshooting. It will also keep you from making dumb mistakes on those pesky finals in school! Learn to estimate in your head as much as possible. It is OK to use calculators and other tools—just keep a running estimation in your head to check your work.

When you are estimating, you are trying to simplify the process of getting to the answer by allowing a margin of error to creep in. The estimated answer you get will be "almost" right, and close enough to help you figure out where else you may have screwed up.

In horseshoes you get a few points for "almost" getting a ringer, but I doubt your boss will be happy with a circuit that "almost" works. However, if your estimates are "almost" right, they can help you design a circuit that even my dad would think is good enough.

Thumb Rules

- Always consider units in your equations; they can help you make sure you are getting the right answer.
- Use units to create the right equation to solve the problem. Do this by making a unit equation and canceling units until you have the result you want.
- Use estimation to determine approximately what the answer should be as you are analyzing and troubleshooting; then compare that to the results to identify mistakes.

How to Visualize Electrical Components

Mechanical engineers have it easy. They can see what they are working on most of the time. As an EE, you do not usually have that luxury. You have to imagine how those pesky electrons are flittering around in your circuit. We are going to cover some basic comparisons that use things you are familiar with to create an intuitive understanding of a circuit. As a side benefit, you will be able to hold your own in a mechanical discussion as well. There are several reasons to do this:

- The typical person understands the physical world more intuitively than they understand the electrical one. This is because we interact with it using all of our senses, whereas the electrical world is still very magical, even to an educated engineer. This is because much of what happens inside a circuit cannot be seen, felt, or heard. Think about it. You flip on a light switch and the light goes on. You really don't consider how the electricity caused it to happen. Drag a heavy box across the floor, and you certainly understand the principle of friction.

- The rules for both disciplines are exactly the same. Once you understand one, you will understand the other. This is great, because you only have to learn the principles once. In the world of Darren we call EEs "sparkies" and ME's "wrenches." If you grok[1] this lesson, a "sparky" can hold his own with the best "wrench" around, and vice versa.

- When you get a feel for what is happening inside a circuit, you can be an amazingly accurate troubleshooter. The human mind is an incredible instrument for simulation and, unlike a computer, it can make intuitive leaps to correct conclusions based on incomplete information. I believe that by learning these similarities you increase your mind's ability to put together clues to the operation and results of a given system, resulting in correct analysis. This will help your mind to "simulate" a circuit.

[1] If you do not know what "grok" means, I highly suggest reading Robert Heinlein's Stranger in a Strange Land. I personally rank it as one of the best 10 books ever written.

Physical Equivalents of Electrical Components

Before we move on to the physical equivalents, let's understand voltage, current and power. Voltage is the potential of the electron flow. Current is the amount of flow. Sometimes the best analogies are the old overused ones and that is true in this case. Think of it in terms of water in a squirt gun. Voltage is the amount of pressure in the gun. Pressure determines how far the water squirts, but a little pea shooter with a 30-foot shot and a dinky little stream won't get you soaked. Current is the size of the water stream, but a large stream that doesn't shoot far is not much help in a water fight. What you need is a supersoaker 29 gagillion, with a half-inch water stream that shoots 30 feet. Now that would be a powerful water-drenching weapon. Voltage, current and power in electrical terms are related the same way. It is in fact a simple relationship; here is the equation:

$$voltage * current = power \qquad \text{Eq. 1-1}$$

To get power you need both voltage and current. If either one of these are zero you get zero power output. Let's discuss three basic components and look at how they relate to voltage and current.

The Resistor Is Analogous to Friction

Think about what happens when you drag a heavy. box across the floor. A force called friction resists the movement of the box. This friction is related to the speed of the box. The faster you try to move the box, the more the friction resists your work. It can be described by an equation.

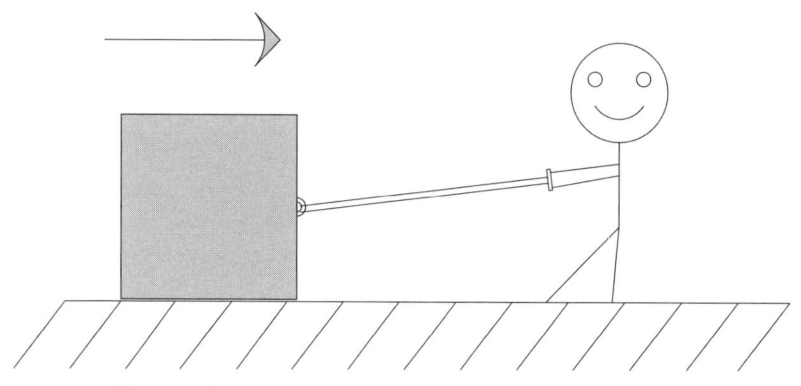

Figure 1-1 *Friction resists smiley stick boy's efforts*

$$friction = \frac{force}{speed} \quad \text{Eq. 1-2}$$

Furthermore, the friction dissipates the energy loss in the system with heat. Let me rephrase that. Friction makes things get warm. Don't believe me? Try rubbing your hands together right now. Did you feel the heat? That is caused by friction. The function of a resistor in an electrical circuit is equal to friction. The resistor resists the flow of electricity just like friction resists the speed of the box. And, guess what, it heats up as it does so. An equation called Ohm's Law describes this relationship:

$$resistance = \frac{voltage}{current} \quad \text{Eq. 1-3}$$

Do you see the similarity to the friction equation? They are exactly the same. The only real difference is the units you are working in.

The Inductor Is Analogous to Mass

Let's stay with the box example for now. First let's eliminate friction, so as not to cloud our comprehension. The box is on a smooth track with virtually frictionless wheels. You notice that it takes some work to get the box going, but once moving, it coasts along nicely. In fact, it takes work to get it to stop again. How much work depends on how heavy the box is. This is known as the law of inertia. Newton postulated this long before electricity was discovered, but it applies very well to inductance. Mass resists a change in speed. Correspondingly, inductance resists a change in current.

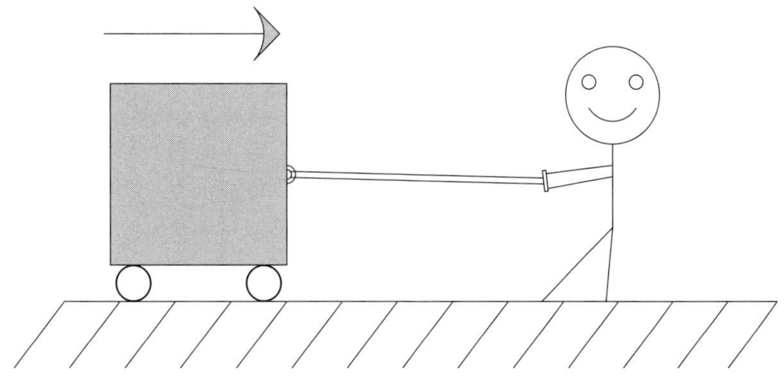

Figure 1-2 *Wheels eliminate friction but smiley has a hard time getting it up to speed and stopping it*

$$mass = \frac{force * time}{speed} \qquad \text{Eq. 1-4}$$

$$inductance = \frac{voltage * time}{current} \qquad \text{Eq. 1-5}$$

The Capacitor Is Analogous to a Spring

So what does a spring do? Take hold of a spring in your mind's eye. Stretch it out and hold it, and then let it go. What happens? It snaps back into position. A spring has a capacity to store energy. When a force is applied, it will hold that energy till it is released. Capacitance is similar to the elasticity of the spring. (One note: the spring constant that you may remember from physics texts is the inverse of the elasticity.) I always thought it was nice that the word capacitor is used to represent a component that has the capacity to store energy.[2]

$$spring = \frac{speed * time}{force} \qquad \text{Eq. 1-6}$$

$$capacitance = \frac{current * time}{force} \qquad \text{Eq. 1-7}$$

A Tank Circuit

Take the basic tank or LC circuit. What does it do? It oscillates. A perfect circuit would go on forever at the resonant frequency. How should this appear in our mechanical circuit? Think about the equivalents: an inductor and a capacitor, a spring and mass. In a thought experiment, hook the spring up to the box from the previous drawing. Now give it a tug. What happens? It oscillates.

---

[2] Technically an inductor can store energy too. In a capacitor the energy is stored in the electric field that is generated in and around the cap; in an inductor energy is stored in the magnetic field that is generated. This energy stored in an inductor can be tapped very efficiently at high currents. That is why most switching power supplies have an inductor in them as the primary passive component.

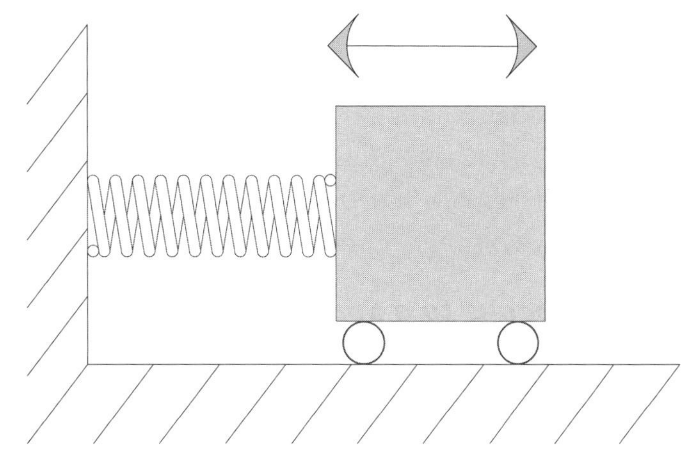

Figure 1-3 *Get this started and it will keep bouncing until friction brings it to a halt*

A Complex Circuit

Let's follow this reasoning for an LCR circuit. All we need to do is add a little resistance, or friction, to the mass-spring of the tank circuit. Let's tighten the wheels on our box a little too much so that they rub. What will happen after you give the box a tug? It will bounce back and forth a bit till it comes to a stop. The friction in the wheels slows it down. This friction component is called a damper, because it dampens the oscillation. What is it that a resistor does to an LC circuit? It dampens the oscillation.

There you have it, the world of electricity reduced to everyday items. Since these components are so similar, all the math tricks you may have learned apply as well to one system as they do to the other. Remember Fourier's theorems? They were discovered for mechanical systems long before anyone realized that they work for electrical circuits as well. Remember all that higher math you used to know or are just now learning about—Laplace transforms, integrals, derivatives, etc.? It all works the same in both worlds. You can solve a mechanical system using Laplace methods just the same as an electrical circuit.

Back in the '50s and '60s, the government spent mounds of dough using electrical circuits to model physical systems as described above. Why? You can get into all sorts of integrals, derivatives and other ugly math when modeling real-world systems. All that can get jumbled quickly after a couple of orders of complexity. Think about an artillery shell fired from a tank. How do you predict where it will land? You have the friction of the air, the mass of the shell, the spring of the recoil. Instead of trying to calculate all that math by hand, you can build a circuit with all the various electrical components representing the mechanical ones, hook up an oscilloscope, and fire away. If you want to test 1000 different weights of artillery at different altitudes, electrons are much cheaper than gunpowder.

Thumb Rules

It takes voltage and current to make power.

A resistor is like friction, it creates heat from current flow (resisting it), proportional to voltage across it.

An inductor is like a mass.

A capacitor is like a spring.

The inductor is the inverse of the capacitor.

Learn an Intuitive Approach

Intuitive Signal Analysis

I'm not sure if this is actually taught in school. This is my name for it. It is something I learned on my own in college and the workplace. I didn't call it an actual discipline until I had been working for a while and had explained my methods to fellow engineers to help them solve their own dilemmas. I do think, however, that a lot of so-called bright people out there use this skill without really knowing it or putting a name to it. They seem to be able to point to something you have been working on for hours and say your problem is there. They just seem to intuitively know what should happen. I believe this is a skill that can and should be taught.

There are three underlying principles needed to apply intuitive signal analysis. (Let's just call it ISA. (After all, if I have any hope of this catching on in the engineering world, it has to have an acronym!)

First, you must drill the basics. For example, what happens to the impedance of a capacitor as frequency increases? It goes down. You should know that type of information off the top of your head. If you do, you can identify a high-pass or low-pass filter immediately. How about the impedance of an inductor—what does it do as frequency increases? What does negative feedback do to an op-amp, how does its output change? You do not necessarily need to know every equation by heart, but you do need to know direction of the change. As far as the magnitude of the change is concerned, if you have a general idea of the strength of the signal that is usually enough to zero in on the part of the circuit that is not doing what you want it to.

Second, you need experience and lots of it. You need to get a feel for how different components work. You need to spend a lot of time in the lab and you need to understand the basics of each component. You need to know what a given signal will do as it passes through a given component. Remember the physical equivalents of the basic components? These are the building blocks of your ability to visualize the operation of a circuit. You must imagine what is happening inside the circuit as the input changes. If you can visualize that, you can predict what the outputs will do.

Third, break the problem down. "How do you eat an elephant?" the knowledge seeker asked the wise old man. "One bite at a time," he replied. Pick a point to start and walk through it. Take the circuit and break it down into smaller chunks that can be handled easily. Draw arrows step by step that show the changes of signals in the circuit. "Does current go up here?" "Voltage at such and such point should be going down." These are the types of questions and answers you should be mumbling to yourself.[3] Again, one thing you do not need to know is what the output will be precisely. You do not need to memorize every equation in the book to intuitively know your circuit, but you do need to know what effect changing a value of a component will have. For example, given a low-pass RC filter and an AC signal input, if you increase the value of the capacitor, what should happen to the amplitude of the output? Will it get smaller or larger? You should know immediately with something this basic that the answer is "smaller." You should also know that how much smaller depends on the frequency of the signal and the time constant of the filter. What happens as you increase current into the base of a transistor? Current through the collector increases. What happens to voltage across a resistor as current decreases? These are simple effects of components, but you would be surprised at how many engineers don't know the answers to these types of questions off the top of their head.

---

[3] Based on extensive research of talking to two or three people, I have concluded that all intelligent people talk to themselves. Whether or not they are considered socially acceptable depends on the audibility of this voice to others around them.

Spending a lot of time in the lab will help immensely to develop this skill. If you look at the response of a lot of different circuits many, many times, you will learn how they should act. When this knowledge is integrated, a wonderful thing happens. Your head becomes a circuit simulator. You will be able to sum up the effects caused by the various components in the circuit and intuitively understand what is happening. Let me show you an example.

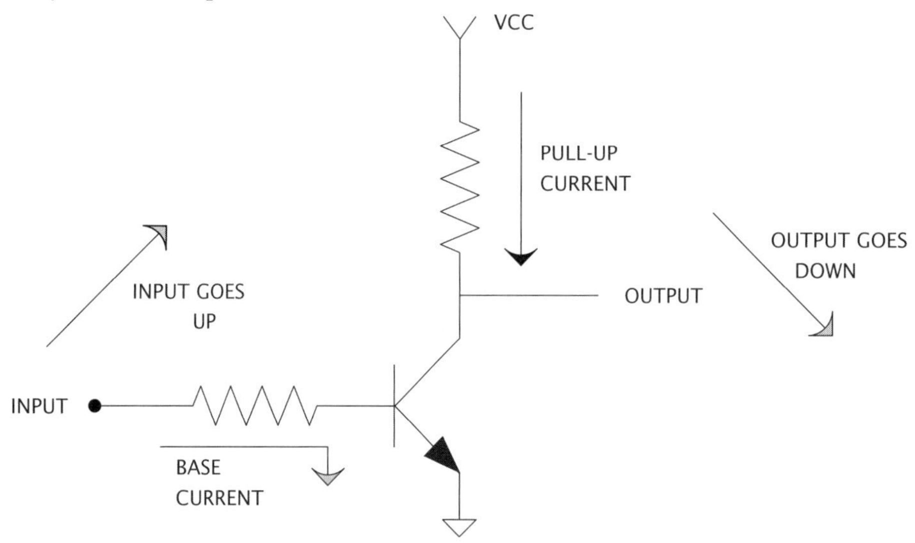

Figure 1-4 *Use arrows to visualize what is happening to voltage and current*

Now at this time you may not have a clue as to what a transistor is, so you may need to file this example away until you get past the transistor chapter, but be sure to come back to it so that the "Aha!" light bulb clicks on over your head. The analysis idea is what I am trying to get across, as you need it early on, but it creates a type of chicken-and-egg dilemma when it comes to an example. So for now, consider this example with the knowledge that the transistor is a device that moves current through the output that is proportional to the current through the base.

As voltage at the input above increases, base current increases. This causes the pull-up current to increase, resulting in a larger voltage drop across the pull-up resistor. This means the voltage at the output must go *down* as the voltage at the input goes *up*. That is an example of putting it all together to really understand how a circuit works.

One way to develop this intuitive understanding is by use of computer simulators. It is easy to change a value and see what effect it has on the output and you can try several different configurations in a short amount of time. However, you have to be careful with these tools. It is easy to fall into a common trap, trusting the simulator so much that you will think there is something wrong with the real world when it doesn't work right in the lab. The real world is not at fault! It is the simulator that is missing something. I think it is best for the engineer to begin using simulators to model simple circuits. Don't jump into a complex model until you grasp what the basic components do—for example, modeling a step input into a RC circuit. With a simple model like this, change the values of R and C to see what happens. This is one way an engineer can develop the correct intuitive understanding of these two components. One word of warning though—don't spend all your time on the simulator. Make sure you get some good bench time too.

You will find this signal analysis skill very useful in diagnosing problems as well as in your design efforts. As your intuitive understanding increases, you will be able to leap to correct conclusions without all the necessary facts. You will know when you are modeling something incorrectly because the result just doesn't look right. Intuition is a skill no computer has, so make sure you take advantage of it!

Thumb Rules

- Drill the basics; know the basic formulas by heart.
- Get a lot of experience with basic circuits; the goal is to intuitively know how a signal will be affected by a component. Spend a lot of time on the bench!
- Break the problem down; draw arrows and notes on the schematic that indicate what the signal is doing.
- Determine what direction the signal is going; is it inversely related or directly related?
- Develop estimation abilities.
- Spend time on the bench with a scope and simple components.

"Lego" Engineering

Building Blocks

OK, so I came up with a fourth item.[4] One of my instructors (we will call him Chuck[5]) taught me a secret that I would like to pass on. Almost every discipline is easier to understand than you might think. The secret professors don't want you to know is there are usually about five or six basic principles or equations that lie at the bottom of the pile, so to speak. These fundamentals, once they are grasped, will allow you to derive the rest of the principles or equations in that field. They are like the old simple Legos®; you had five or six shapes to make everything. If you truly understand these few basic fundamentals in a given discipline, you will excel in that discipline. One other thing Chuck often said was that all the great discoveries were only one or two levels above these fundamentals. This means if you really know the basics well, you will excel at the rest. One thing you can be sure of is the human tendency to forget. All the higher level stuff is often left unused and will quickly be forgotten, but even an engineer-turned-manager like me uses the basics nearly every day.

Since this is a book on electrical engineering, let's list the fundamental equations for electrical circuits as I see them:

- Ohm's Law
- Voltage divider rule
- Capacitors impede changes in voltage
- Inductors impede changes in current
- Series and parallel resistors
- Thevenin's theorem

We will get into these in more detail later in the chapters, but let me touch on a couple of examples. You may say, "You didn't even list series and parallel capacitors. Isn't

[4] For those of you that have been wondering if I can count.

[5] Dr. Charles Tinney was what he wrote on the chalkboard the first day of class. Then he turned around and said, "You can call me Chuck!" I have to credit Dr. Tinney—he was the best teacher I have ever had. For him nothing was impossible to understand, or to teach you to understand.

that a basic rule?" Well, you are right, it is fairly basic but it really isn't at the bottom of the pile. Series and parallel resistors are even more fundamental, because all that really happens when you add in the caps is the frequency of the signal is taken into account, other than that it is exactly the same equation! You would be better served to understand how a capacitor or inductor works and apply it to the basics than try to memorize too many equations. "What about Norton's theorem?" you might ask. Bottom line, it is just the flip side of Thevenin's theorem, so why learn two when one will do? I prefer to think of it terms of voltage so I set this to memory. You could work in terms of current and use Norton's theorem, but you would arrive at the same answer at the end of the day. So pick one and go with it.

You can always look the more advanced stuff up, but most of the time a solid application of the basics will force the problem at hand to submit to your engineering prowess. These six rules are things that you should memorize, understand and be able to do approximations of in your head. These are the rules that will make the intuition you are developing a powerful tool. It will unleash the simulation capability that you have right in your own brain.

If you really take this advice to heart, years down the road when you've been given your "pointy hairs"[6] and you have forgotten all the advanced stuff you used to know you will still be able to solve engineering problems to the amazement of your engineers.

This can be generalized to all disciplines. Look at what you are trying to learn, figure out the few basic points being made from which you can derive the rest, and you will have discovered the basic "Legos" for that subject. Those are what you should know forwards and backwards to succeed in that field. Besides, Legos are fun, aren't they?

Thumb Rules

- There are a few rules in any discipline from which you can derive the rest.
- Learn these rules by heart; gain an intuitive understanding of them.
- Most significant discoveries are only a level or two above these basics.

[6] In case you have lived under a rock for the last few years and missed a certain very successful engineering cartoon, this means promoted to management.

CHAPTER 2

Basic Theory

Every discipline has fundamentals that are used to extrapolate all the other more complex ideas. Basics are the most important thing you can know. It is knowledge of the basics that helps you apply all that stuff in your head correctly. It doesn't matter if you can handle quadratic equations and calculus in your sleep. If you don't grasp the basics, you will find yourself constantly chasing a problem in circles without resolution. If you get anything out of this text, make sure you really understand the basics!

Ohm's Law Still Works

Ohm's Law

This, I believe, is one of the best-taught principles in school, as it should be. So why go over it? Well, two reasons come to mind: one, you can't go over the basics too much, and two, while any engineer can quote Ohm's Law by heart, I have often seen it ignored in application.

First let's state Ohm's Law: voltage equals current multiplied by resistance.

$$V = I * R$$

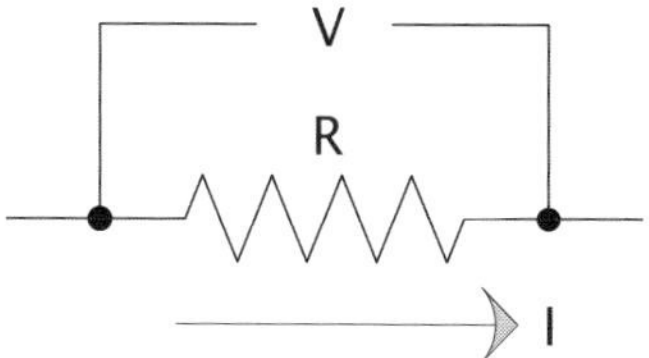

V
R
I

$$V = I \cdot R$$

$$I = \frac{V}{R}$$

$$R = \frac{V}{I}$$

Figure 2-1 *Ohm's Law, the heart of all things electrical*

It is simple, but do you consider that resistance exists in every part of a circuit? (Unless it is a superconductor.) It is easy to forget that, especially since many simulators do. I think the best way to drive this point home is to recount the way it was driven home to me.

There I was, a lowly engineering student. I was working as a technician, or associate engineer (it depended on who you asked). I was arguing with my boss who had an MSEE degree, but he just wouldn't believe me; neither would my lead engineer (who had a BSEE). I couldn't bring myself to distrust Ohm's Law, however, even in light of their "superior" knowledge. I'd had less heated debates with rabid dogs. This was the problem. Our department needed to measure the current of a DC motor that could range from 5A to 15A at any given time, but our multimeters had a 10A fuse in the current measuring circuit. So, using Ohm's Law (which was fresh in my mind being a student and all), I designed a shunt to measure current. I wanted to get a good reading but disturb the circuit as little as possible, so I chose a 0.1Ω resistor. I built a box to house it and installed banana jack plugs to provide an easy interface to a voltmeter. The design looked like this:

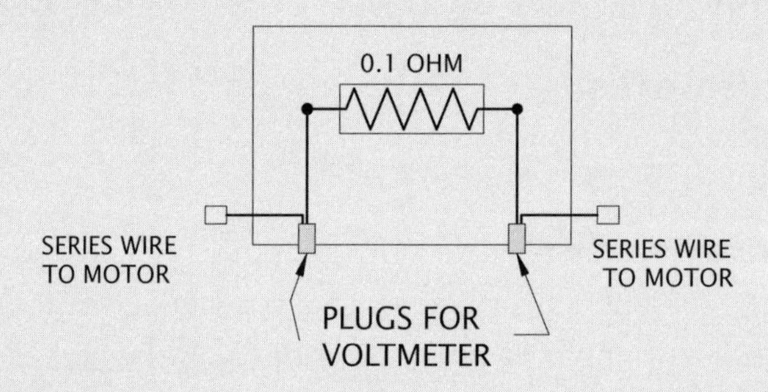

Figure 2-2 *Original design of simple current-measuring circuit*

Everyone thought it was a great idea so I built a couple of boxes and we started using them right away. After a while, however, we noticed that they were not very accurate. Sometimes they would be off by as much as 50% to 60%. No one could figure out why so I sat down to analyze what I had created.

After a few minutes, I said to myself, "Well, duhhh!" I realized that to make the assembly easy I had soldered the wires from the motor to the banana jacks and then soldered some short 14-gauge jumpers to the shunt resistor. My circuit really looked like this:

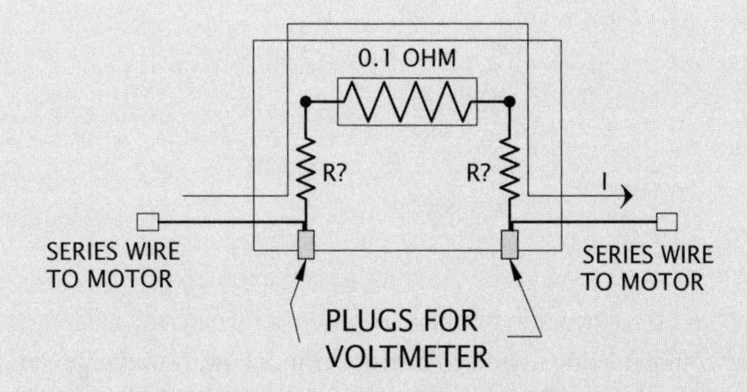

Figure 2-3 *As-built simple current-measuring circuit*

My voltmeter was measuring across a larger resistance value than 0.1 ohms. Wire has resistance too; even a couple of inches of 14-gauge wire has a few hundredths of an ohm. Remembering Ohm's Law…

$$V = I * R$$

I realized this means if you increase R, you get more V for the same amount of current, leading to the errors we were seeing. I had made a simple mistake that fortunately was easy to correct. I redesigned the box on paper to look like the following:

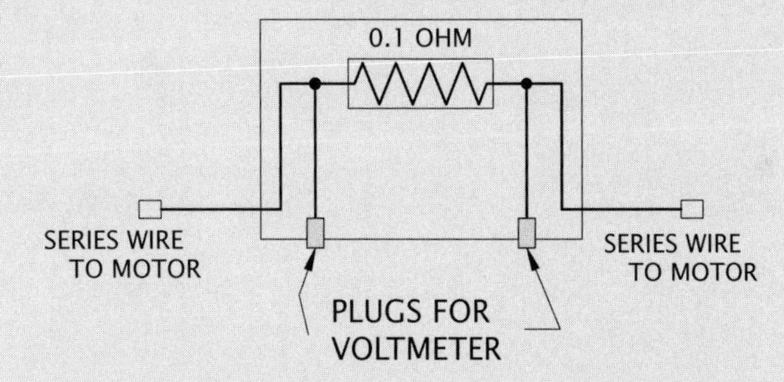

Figure 2-4 *Redesigned current-measuring circuit*

I took this to my boss (the one with the MSEE that could do math in his head that I would only attempt with MathCad and a cold drink). His reaction floored me. He reviewed it with the lead engineer and they came to the conclusion that I was completely wrong. They were talking about things like temperature coefficients and phase shifts in current and RMS and a bunch of other topics that were over my head at the time. Thus began the argument. I explained that two points on a schematic had to be connected by a wire and a wire had resistance. While it is often ignored, it was significant in this case because the shunt resistor was such a small value.

As they hemmed and hawed over this, I learned that many times it is human nature to ignore what one learned long ago and try to apply more advanced theories just because you know them. Also, all the knowledge in the world isn't worth jack if it is incorrectly applied. I continued to press my point. I must have written Ohm's Law on the white board fifty times by then.

They finally conceded and agreed that the extra wire between the banana jack and the shunt was the cause of the error. But that was not the end of the disagreement. How in the world was my new design going to fix the problem by simply repositioning the wires? The resistance was still in the circuit, was it not? I wrote down Ohm's Law another hundred times, and explained that the current through the meter was very small, making the resistance in the wire insignificant again. My astonishment reached new levels as I observed the human ability to overlook the obvious. The first argument was nothing compared to this one. The fireworks started to really fly then.

What is the moral of this story? Well, Scott Adams, creator of Dilbert said, "Everyone has moments of stupidity" as he watched someone fix his "broken" pager by putting in a new battery. I have to agree with him. I rediscover Ohm's Law about every 6 months. Always, always, always check the basics before you start looking for more complicated solutions! My father, a mechanic, tells a story of rewiring an entire car just to find a bad fuse. (It looked okay, but didn't check out with a meter.) That was how he learned this lesson. Me, I just participated in 4 hours of the dumbest argument of my career.

How did the argument end? We never came to an agreement, so I went ahead and fixed boxes with the new design anyway (which they spent several weeks proving were working correctly). I didn't say another word, but transferred out of that group as soon as possible. The same design has been in use for over 10 years now, and the documentation notes the need to wire it correctly to avoid inaccurate readings. I didn't write that document, my old boss did. It's kind of funny how we didn't argue about Ohm's Law after that.

The basics are the most important; let me repeat that, the BASICS are IMPORTANT! Ohm's Law is the most basic principle you will use as an electrical engineer. It is the foundation on which all other rules are based. The fundamental fact is that resistance impedes current flow. This impedance creates a voltage drop across the resistor that is proportional to the amount of current flowing through it. If it helps, you can think of a resistor as a current-to-voltage converter.

With that important point made, let's consider two other types of impedance that can be found in a circuit. We will get into this in more detail later, but for now consider that inductors and capacitors both can act like resistors depending on the frequency of the signal. If you take this into account, Ohm's Law still works when applied to these components as well. You could very well rewrite the equation to:

$$V = I * Z \qquad \text{Eq. 2-1}$$

Think of the impedance Z as a resistance at a given frequency. As we move on, keep this in mind. Wherever you see resistance in an equation you can simply replace it with impedance if you consider the frequency of the signal. (Not a perfect analogy, but it will help you intuitively get the idea of impedance.)

One final note before we move on. Every wire, trace, component, or material in your circuit has these three components in it: resistance, inductance, capacitance. Everything has resistance, everything has capacitance, everything has inductance, and the question is, "Is it enough to make a difference?" The fact is, in my own experience, if the shunt resistor had been 100 times larger, that would have made the errors we were seeing 100 times less. They would have been insignificant in comparison to the measurement we were taking. The impedance equations for capacitors and inductors will help you similarly consider the frequencies you are operating at and ask yourself, "Is this component making a significant impact on what I am looking at?" By reviewing this significance, you will be able to pinpoint the part of the circuit you are looking for.

The experience I related earlier happened years ago at the beginning of my career, and I said then that I still rediscover Ohm's Law every six months. Time and time again, working through a problem or design, the answer can be found by application of Ohm's Law. So before you break out all those higher theories trying to solve a problem, first remember: Ohm's Law still works!

The Voltage Divider Rule

$$Vo = Vi \frac{Rg}{Rg + Ri} \qquad \text{Eq. 2-2}$$

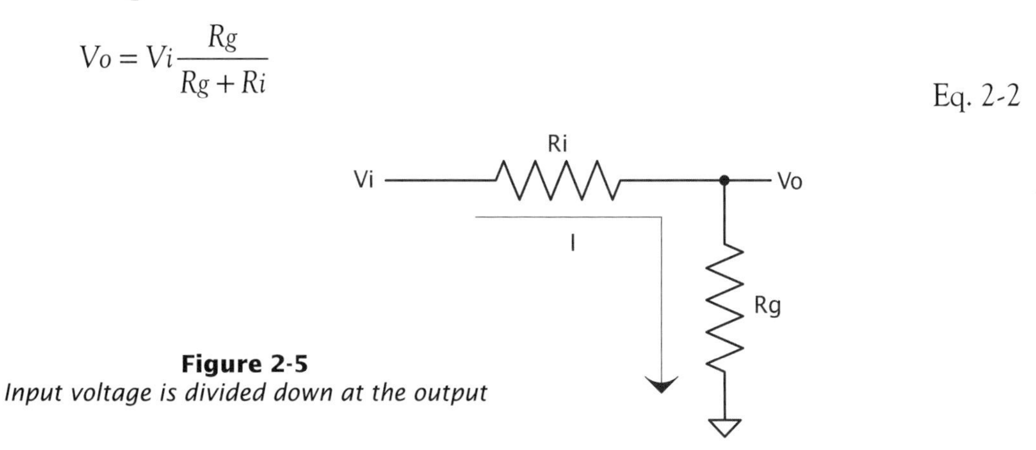

Figure 2-5
Input voltage is divided down at the output

The most common way you will see this is in terms of R1 and R2. I have changed these to *Rg* (for *R* ground) and *Ri* (for *R* input) to remind myself which one of these goes to ground and which one is in series. If you get them backwards you get the amount of voltage lost across *Ri*, not the amount at the output (which is the voltage across *Rg*). If the gain of this circuit[1] just doesn't seem right, you may have the two values swapped.

You may also notice that the gain of this circuit is never greater than one. It approaches one as *Ri* goes to zero, and it approaches one as *Rg* gets very large. (Note as *Rg* gets larger the value of *Ri* becomes less significant.) Since this is the case, it is easy to think of the voltage divider as a circuit that passes a percentage of the voltage through to the output. When you look at this circuit, try to think of it in terms of percentage. For example, if *Rg* = *Ri*, only 50% of the voltage would be present on the output. If you want 10% of the signal, you will need a gain of 1/10. So put 1K in for Rg, and 9K in for *Ri*, and, voila, you have a voltage divider that dissipates 90% of the signal as heat across the resistors, leaving you 10% of the signal at the output.

Did you notice that the ratio of the resistors to each other was 1:9 for a gain of 1/10? This is because the denominator is the sum of the two resistor values. I'll also bet you noticed that if you swap the two resistor values you will get a gain of 9/10 or 90%. This should make intuitive sense to you now if you recognize that, for the same amount of current, the voltage drop across a 9K *Ri* will be 9 times larger than the voltage drop across a 1K *Rg*. In other words, 90% of the voltage is across *Ri*, while 10% of the voltage is across *Rg*, where your meter measuring Vo is hooked up. The voltage divider is really just an extension of Ohm's Law (go figure) but it is so useful that I've included it as one of the basic equations that you should commit to memory.

[1] One way I like to think of this is $Vo = Vi * H$ where H is the gain of the circuit or $H = Rg/(Rg + Rs)$. This is useful when you are breaking a circuit down to components. We will specifically use this when we discuss op-amps later on.

Capacitors Impede Changes in Voltage

Let's consider for a moment what might happen to the previous voltage divider circuit if we replace *Rg* with a capacitor. It is still a voltage divider circuit, is it not? But what is the difference? At this point you should say, "Hey, a cap is just a resistor who's value changes depending on the frequency; wouldn't that make this a voltage divider that depends on frequency?" Well, it does and this is commonly known as an RC circuit. Let's draw one now.

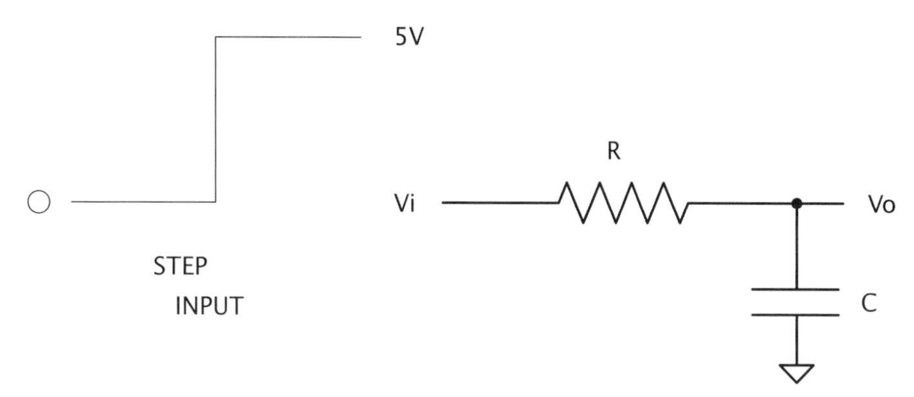

Figure 2-6 *Step input is applied to a simple RC circuit*

Using your intuitive understanding of resistors and capacitors, we will analyze what is going to happen in this circuit. We will apply a step input. A step input is by definition a fast change in voltage. The resistor doesn't care about the change in voltage, but the cap does. This fast change in voltage can be thought of as high frequencies[2], and how does the cap respond to high frequencies? That's right, it has low impedance. So now we apply the voltage divider rule. If the impedance of *Rg* is low (as compared to *Ri*), the voltage at *Vo* is low. As frequency drops, the impedance goes up; as the impedance goes up, based on the voltage divider the output voltage goes up. Where does it all stop? Think about it a moment. Based on what you know about a cap, it resists a change in voltage. A quick change in voltage is what happened initially. After that our step input remained at 5V not changing any more. Doesn't it make sense that the cap will eventually charge to 5V and stay there? This is known

[2] This is something a man named Fourier thought of long ago. The more harmonics you sum together, the faster the rise time of said step input.

as the transient response of an RC circuit. The change in voltage on the output of this circuit has a characteristic curve. It is described by this equation:

$$Vo = Vi\left(1 - e^{-\frac{t}{rc}}\right) \qquad \text{Eq. 2-3}$$

The graph of this output looks like this:

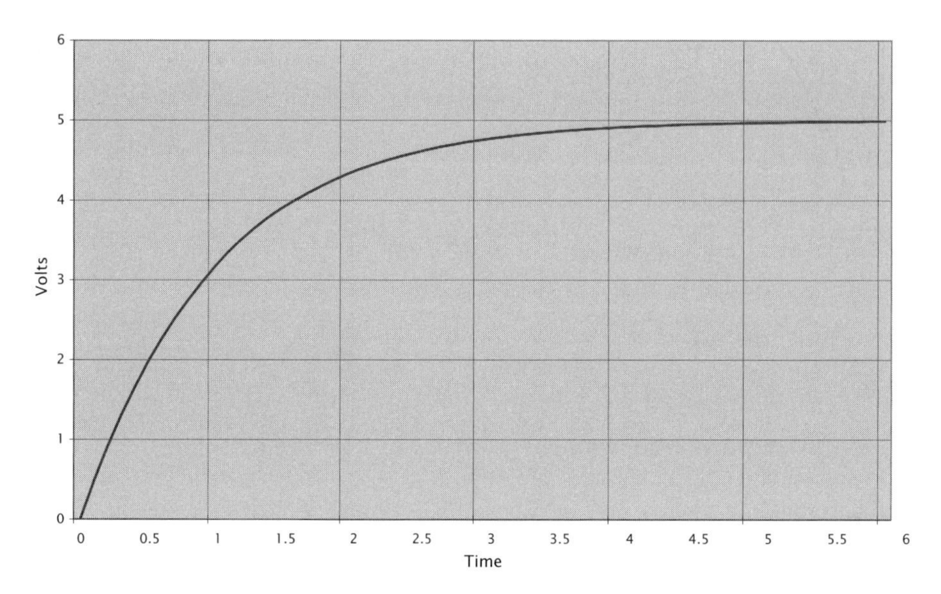

Figure 2-7 *Voltage change over time*

The value of R times C is also known as tau, or the time constant, often referred to by the Greek letter τ.

$$RC = \tau \qquad \text{Eq. 2-4}$$

For a step input, this curve is always the same. The only thing that changes is the amount of time it takes to get to the final value. The shape is the same and the time depends on the value of the time constant tau. You can normalize this curve in terms of the time constant and the final value of the voltage. Let's redraw the curve with multiples of τ along the time axis.

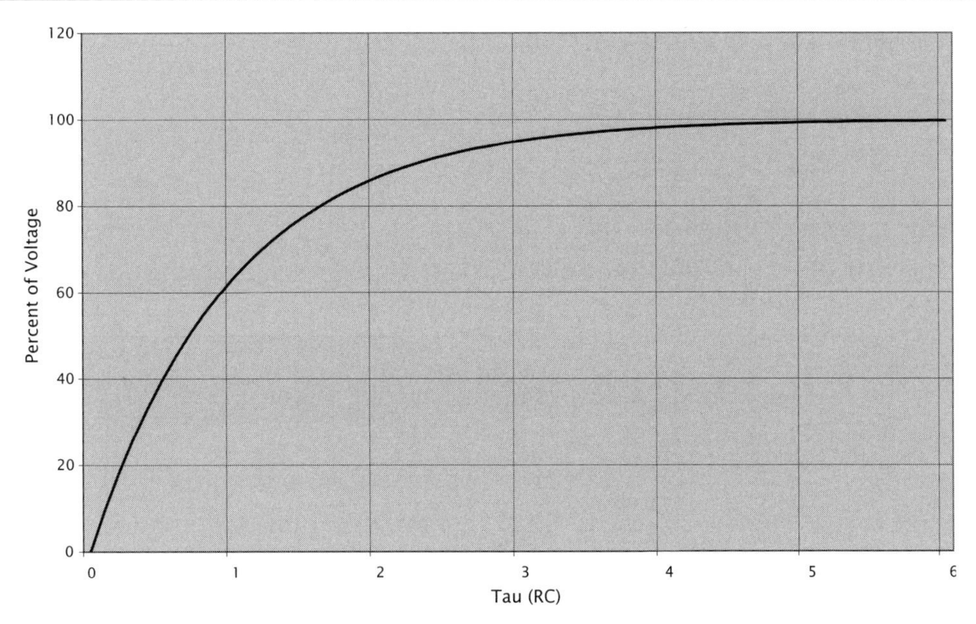

Figure 2-8 *Voltage change in percent over time in tau*

At 1τ the voltage reaches 63.2%, at 2τ it is at 86.6%, 3τ is 95%, by 4τ it is at 98%, and when you reach 5τ you are close enough to 100% to consider it so.

This response curve describes a basic and fundamental principle in electronics. Some years ago I started asking potential job candidates to draw this curve after giving them the RC circuit as in the figure above. Over the years I have been totally dismayed by how many engineers, both fresh out of school and with years of experience, cannot draw this. Less than 50% of the applicants I have asked can do it. That fact is one of the main reasons for writing this book. (The other was that someone was actually willing to pay me to do it! I doubt it would have gotten far otherwise.) So I implore you to put this to memory once and for all; by doing so I guarantee you will be a better engineer! Plus, if I ever interview you, you will have a 50% better chance of getting a job! If you understand this, you will understand inductors, as you will see in the next section.

Before we move on, I would like you to consider what happens to the current in this circuit. Remember Ohm's Law? Apply it to this to understand what the current does. We know that:

$$V = I * R \qquad \text{Eq. 2-5}$$

A little algebra turns this equation into:

$$I = \frac{V}{R}$$ Eq. 2-6

A little common sense reveals that the voltage across R in this circuit is equal to voltage at the output minus voltage at the input. Turn that into an equation and you get:

$$Vr = Vi - Vo$$ Eq. 2-7

We know the voltage at each point in time in terms of tau. At 0τ *Vo* is at 0. So the full 5V is across the resistor and the maximum current is flowing. For all intents and purposes, the cap is shorting the output to ground at this point in time. At 1τ *Vo* is at 63.2% of *Vi*. That means Vr is at 36.8% of *Vi*. Repeat this process, connect the dots and you get a curve that moves in the opposite direction of the voltage curve, something like this:

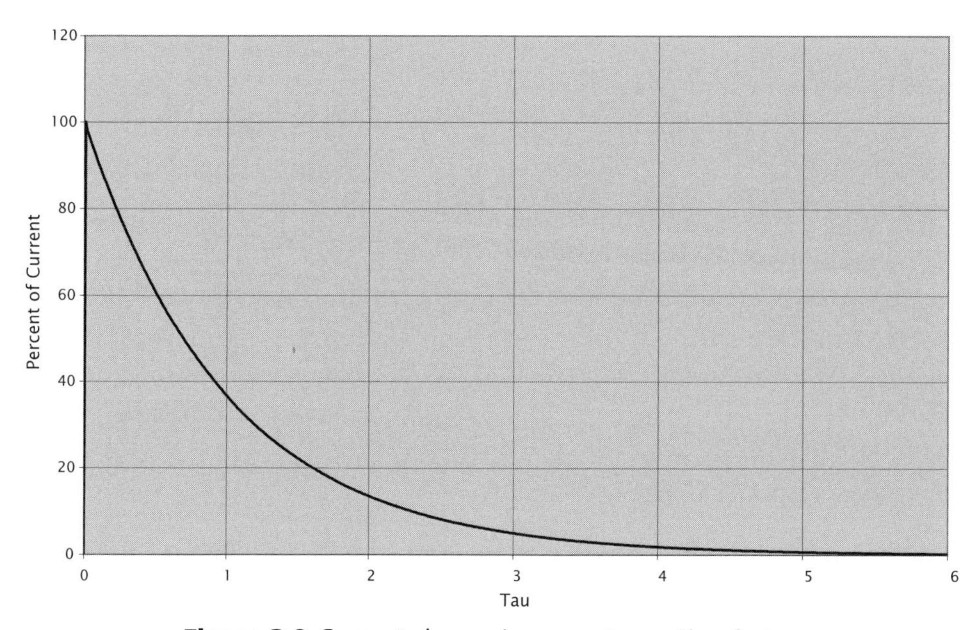

Figure 2-9 *Current change in percent over time in tau*

Notice how current can change immediately when the step input changes? Also notice how the voltage just doesn't change that fast. Capacitors impede a change in voltage as the rule goes. What this also means is that changes in current will not be affected at all. Now everything has its opposite and capacitors are no exception, so let's move on to inductors.

Inductors Impede Changes in Current

Now that we have thought through the RC circuit, let's consider an RL circuit.

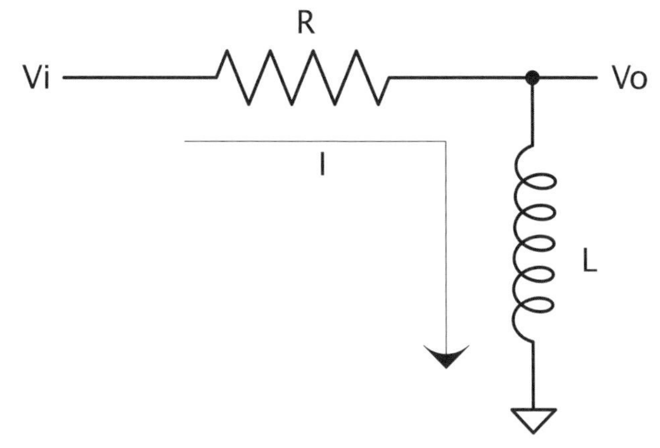

Figure 2-10 *The basic RL circuit*

Remember that the inductor resists a change in current but not in voltage. Initially with the same step input, the voltage at the output can jump right to 5V. Current through the inductor is initially at 0, but now there is a voltage drop across it. So current has to start climbing. The current responds in the RL circuit exactly the same way the voltage responds in the RC circuit.

Since you committed the RC response to memory, the RL response is easy. It is exactly the same from the viewpoint of current; the current graph looks like this:

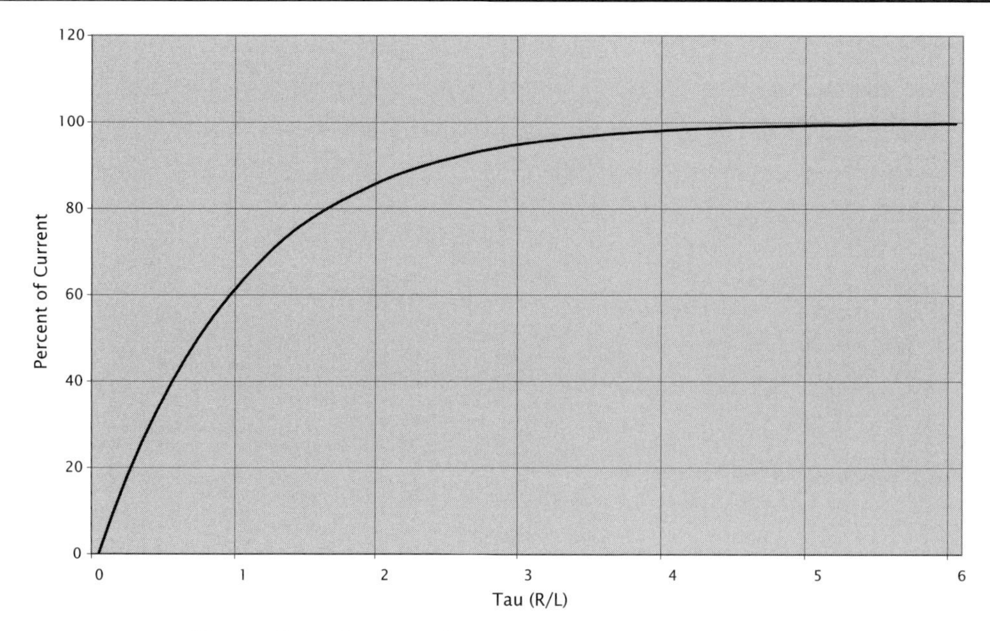

Figure 2-11 *Current change in percent over time in tau*

I hope you are saying to yourself, "What about the voltage response?" At this time I would like you to consider Ohm's Law for a moment and try to graph what the voltage will do. What is the current at time 0? How about a little later? Remember Ohm's Law—for the current to be low, resistance must be high. So initially the inductor acts like an open circuit. Voltage across the inductor will be at the same value as the input. As time goes on, the impedance of the inductor drops off, becoming a short, so voltage drops as well. The graph looks like the following.

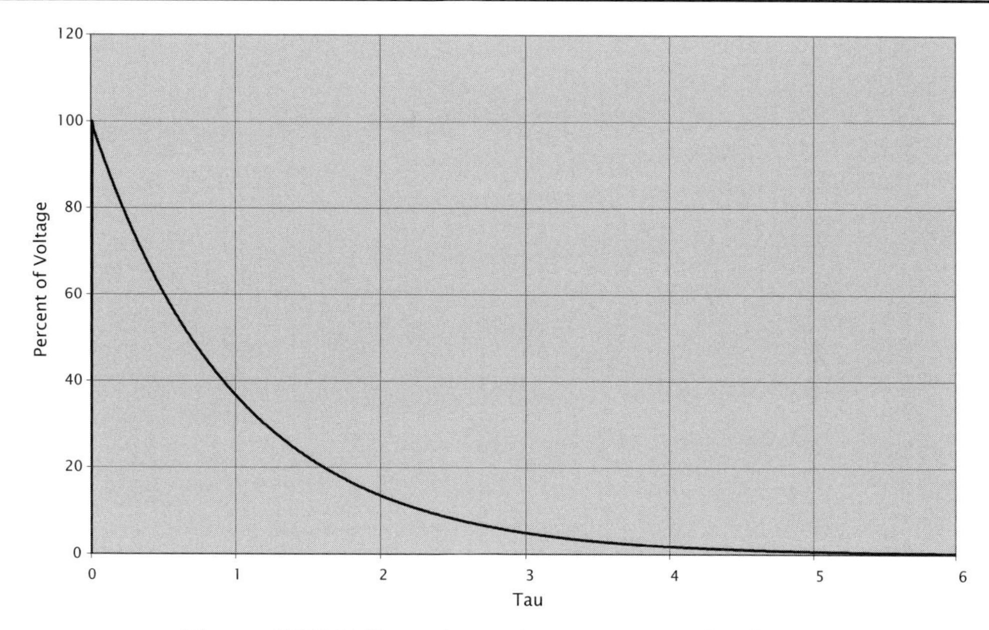

Figure 2-12 *Voltage change in percent over time in tau*

The inductor is the exact complement of the capacitor. What it does to current, the cap does to voltage and vice versa.

Series and Parallel Components

There are two ways for components to be configured in a circuit, series and parallel. Series components line up one after another, while parallel components are hooked up next to each other. Let's go over the formulas to simplify these component arrangements.

Series resistors are easy; you just add them up, no multiplication needed!

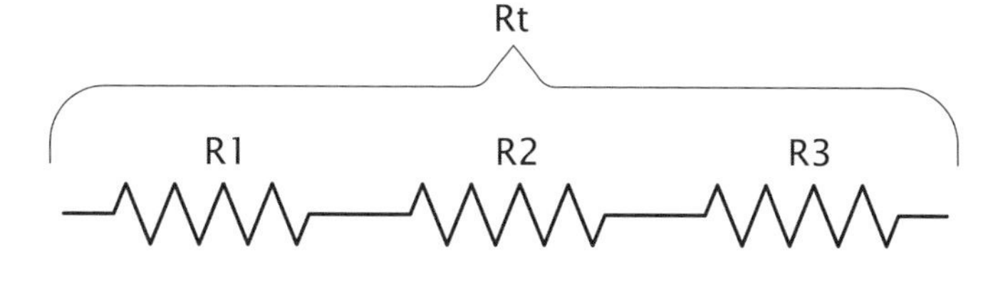

Figure 2-13 *Series resistors*

$$Rt = R1 + R2 + R3 \qquad \text{Eq. 2-8}$$

Inductors are like resistors—you sum series inductors in the same way.

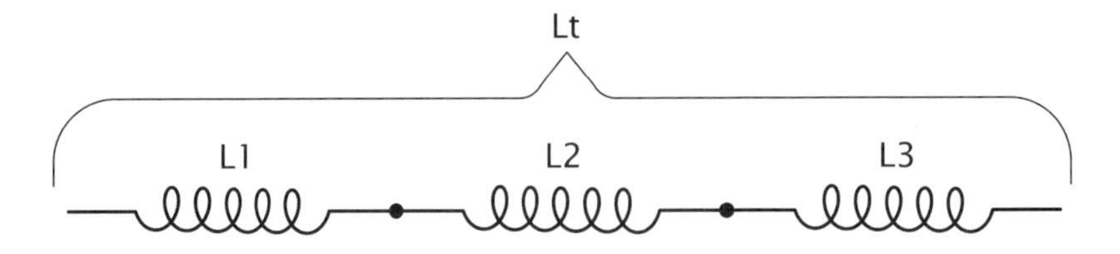

Figure 2-14 *Series inductors*

$$Lt = L1 + L2 + L3 \qquad \text{Eq. 2-9}$$

Remember that capacitors are the opposite of inductors. Due to this, capacitors must be in parallel to be summed up like resistors and inductors are in series.

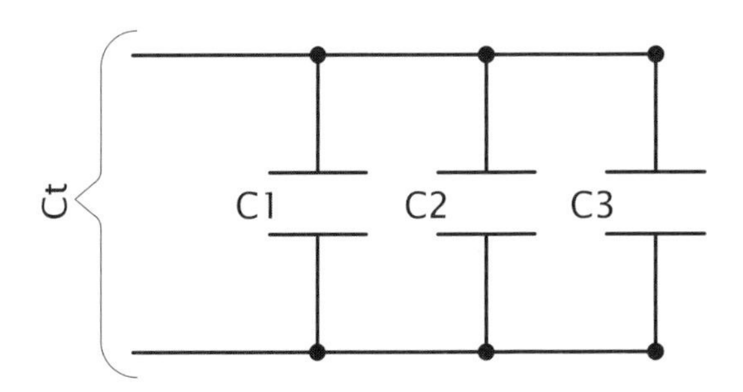

Figure 2-15 *Parallel capacitors*

$$Ct = C1 + C2 + C3 \qquad \text{Eq. 2-10}$$

Remember these equivalences:

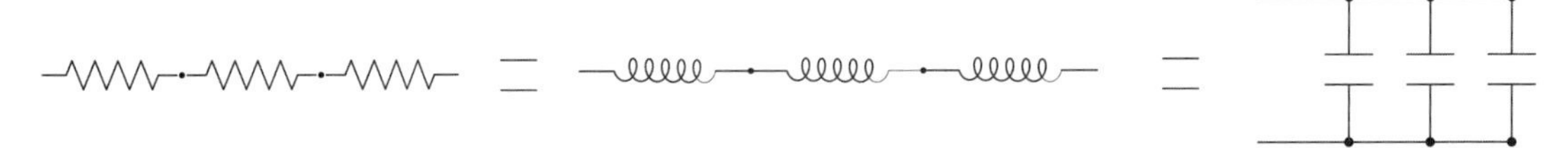

Figure 2-16 *Component equivalents*

Parallel resistors are a little trickier. The equivalent resistance of any two components is determined by the product of the values divided by the sum of the values. Keep in mind though, this works for any **two** resistors! In the case of three resistors, solve any two and repeat until done. (The '//' means in parallel with.)

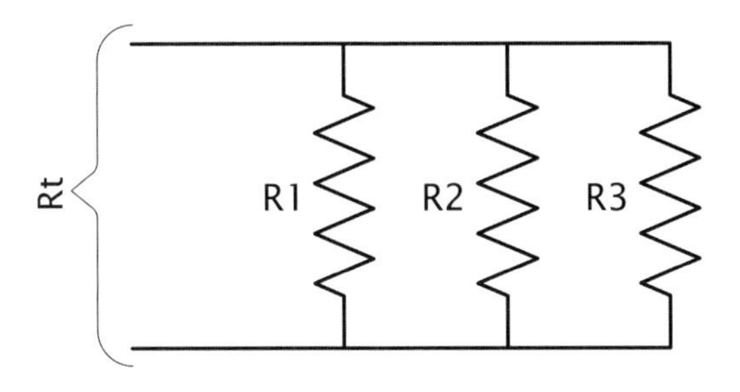

Figure 2-17 *Parallel resistors*

$$R1 // R2 = \frac{R1 * R2}{R1 + R2} \qquad Rt = \frac{R1 // R2 * R3}{R1 // R2 + R3} \qquad \text{Eq. 2-11}$$

Inductors are again the same as resistors. Hooking them up in parallel, you can reduce them the same way:

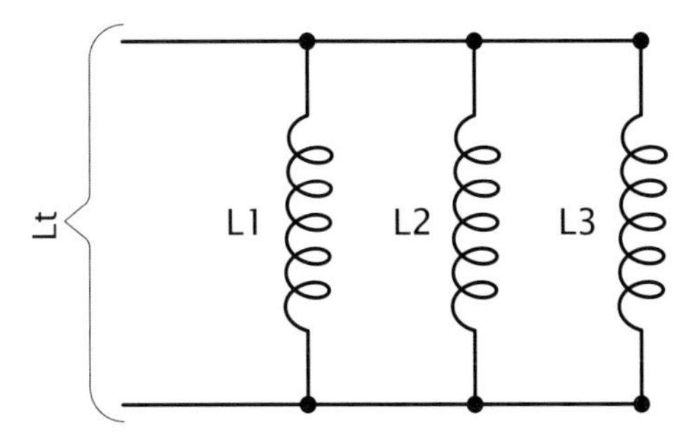

Figure 2-18 *Parallel inductors*

$$L1 // L2 = \frac{L1 * L2}{L1 + L2} \qquad Lt = \frac{L1 // L2 * L3}{L1 // L2 + L3} \qquad \text{Eq. 2-12}$$

For capacitors the same equation applies, but only if they are in series like this:

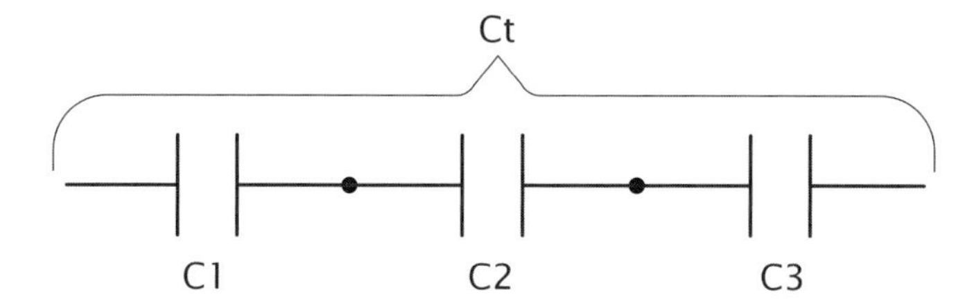

Figure 2-19 *Series capacitors*

$$C1//C2 = \frac{C1 * C2}{C1 + C2} \qquad Ct = \frac{C1//C2 * C3}{C1//C2 + C3} \qquad \text{Eq. 2-13}$$

These are the circuits that use the product over the sum rule.[3]

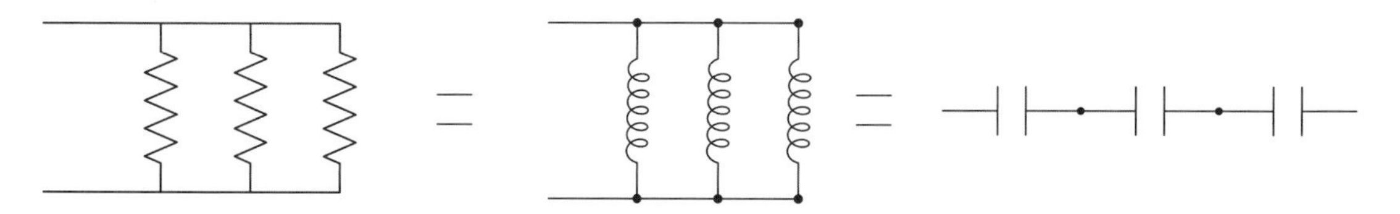

Figure 2-20 *Component equivalents*

In dealing with parallel and series circuits, you can see that there are only two types of equations. One is simple addition, and the other is product over the sum. The only trick is to know which to use when. Remember that the resistor and inductor are part of the "in" crowd and the cap is the outcast wallflower who is the opposite of those other guys. I'll bet most engineers can relate to being the "capacitor" at a party, so this shouldn't be too hard to remember!

[3] Another way to solve this is to take the inverse of the sum of all the inverses ($1/Zt = 1/Z1 + 1/Z2 + 1/Z3$). If this works better for you that is fine, just commit one or the other to memory.

Thevenin's Theorem

Thevenizing is based on the idea of using superposition to analyze a circuit. When you have two different variables affecting an equation, making it difficult to analyze, you can use the technique of superposition to solve the equation provided you are dealing with linear equations (by luck all these basic components are linear, even if you might not think it when looking at the curve of an RC time response, it actually is a linear equation.[4])

The idea of superposition is simple. When you have multiple inputs affecting an output, you can analyze the effects of each input independently and add them together when you are all done to see what the output does. One idea that comes from superposition is Thevenin's Theorem.

Using Thevenin's Theorem allows you to reduce basically any circuit into a voltage divider and we know how to solve a voltage divider, don't we! There is a sister theorem called Norton's, which does the same thing but is based on current rather than voltage. Since you can solve any electrical problem with either equation, I suggest you focus on one or the other. Since I like to think in terms of voltage, I prefer Thevenizing a circuit to the Norton equivalent. So to be true to the idea that you should only learn a few fundamentals and learn those well, we will focus on Thevenin equivalents.

The most important rule when Thevenizing is this: *Voltage sources are shorted, current sources are opened*. Consider the following circuit.

[4] When I see the term *linear equation*, I think line, so an RC curve seems counterintuitive, but linear equations are a type of formula that allows certain rules to be used such as superposition.

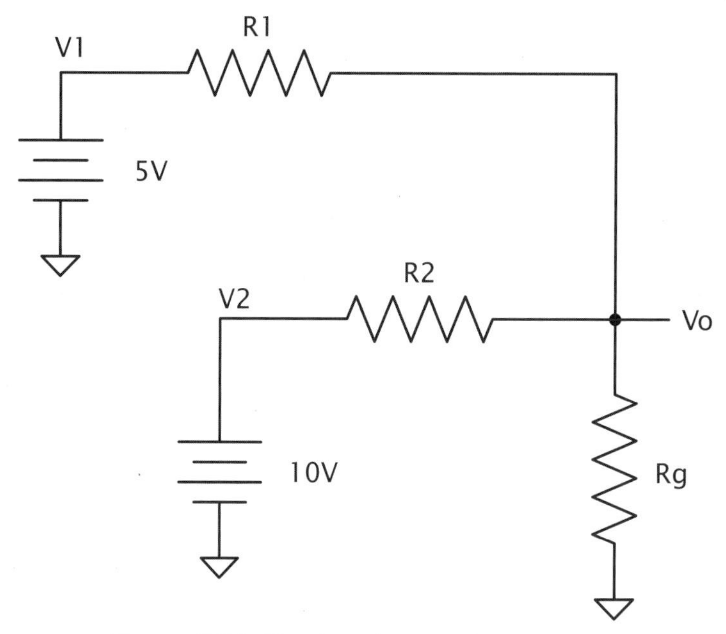

Figure 2-21 *Circuit with two voltage sources*

Once all the voltage sources are shorted and all the current sources are opened, all the components will be in series or parallel. That makes it very convenient for those of us who only want to memorize a few equations! Apply those basic parallel and series rules we just learned and, viola, you have a circuit that is much easier to understand. Once you have reduced the resistors, inductors and caps to a more controllable amount, you replace each source one at a time to see the effects of each source on the component in question.

Once you have considered the effects of each source one at a time, you can add them all together to see the overall effect. In this process, you have Thevenized the circuit and you are superimposing each output on top of the other to get the output of the combined inputs.

It helps when Thevenizing a circuit to try to imagine you are looking back into the circuit from the output. This means you imagine what the circuit looks like in terms of the output. We often think in terms of stuff that goes in the input, something happens, and then it comes out the output. Try flipping that notion on its head. Think here is the output, what exactly is it hooked up to? What are the impedances that the cap in this case "sees" connected to it? Once you are able to adjust your point of view, Thevenizing will become an even more powerful tool. Consider this circuit:

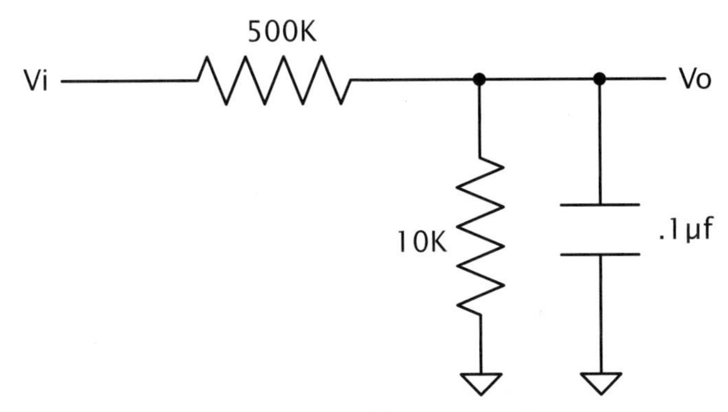

Figure 2-22 *Real live secret circuit*

This circuit comes from a real live application. I would tell you what, but it is secret.[5] So we won't say this is anything more than a voltage divider with a capacitive filter on it. (Values may have been changed to protect the innocent.)

This circuit's job is to lower a voltage at the input terminals varying from 0 to 100 volts to something with a range of 0 to 5 volts. The input voltage also has an AC component that is filtered out by the capacitor. The question is, what is the time constant of the RC filter in this circuit?

Is it 500K * 0.1 µf? That is what I would have thought before I understood Thevenin's Theorem. The output in this case is the voltage across the cap, so let's look back into the circuit to figure out what is hooked up to this cap. Now, remember, we said there was a voltage source on the input of this circuit. Let's short that on our drawing and Thevenize this puppy!

Following is the Thevenized circuit.

[5] If you haven't already, you will soon find out every corporation wants you to sign away every idea you have or have ever had as their intellectual property. Some day, those who have all the good ideas must rise up and say, "enough is enough." After which we will all likely end up being consultants.

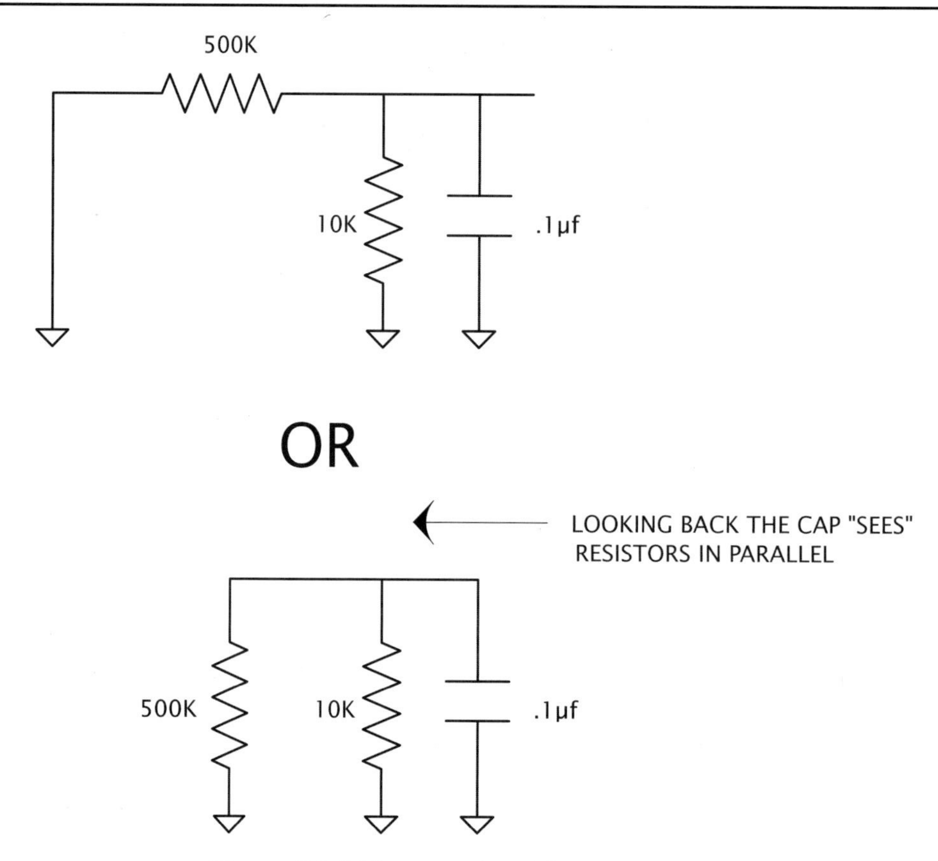

Figure 2-23 *Thevenized real live secret circuit*

Hopefully at this point something really jumped out at you. The 10K and the 500K resistors are in parallel as far as the cap is concerned. Applying the rule of parallel resistors we find that the resistance hooked up to this cap is 9.8K. Wow, that is a lot less than 500K, isn't it! Thevenizing showed us that our first assumption was incorrect. In fact, the time constant[6] of this circuit is much, much lower than it would be without the 10K resistor.

There are other ways this theorem can be useful. Here is a case in point. You might have a circuit like this:

[6] If you want a lesson on time constants, you will need to jump ahead a few chapters, for now though it is sufficient to understand the basic idea behind good o' Thevenin's proposition.

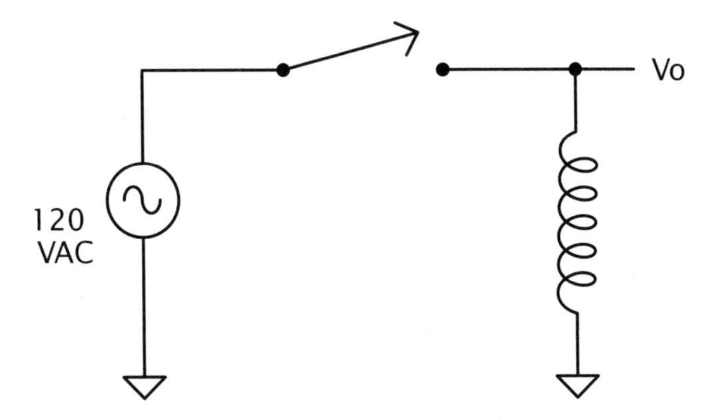

Figure 2-24 *AC switched power to an inductor*

You need to switch AC power through this inductor (which was actually one winding of an AC motor in this case). Trouble is, when you let off the switch a whole bunch of electrical noise is generated when this switch is opened. (We will discuss why when we cover magnetic fields later in the book.) A standard way to deal with this is with an RC circuit commonly known as a snubber. The point of a snubber is to snub this voltage spike and dissipate it as heat on the resistor. This makes the most sense if it is across the inductor, like this:

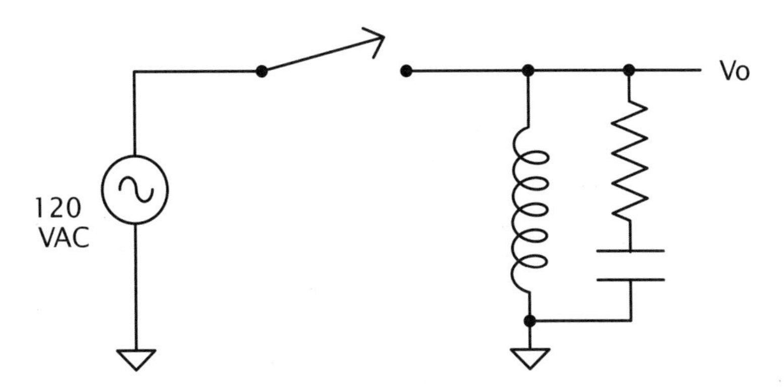

Figure 2-25 *AC switched power to an inductor with snubber*

Now let's apply Thevenin's Theorem to take a different look at this circuit. By shorting the AC voltage source, we quickly see that hooking the snubber up to the other side of the switch, to the AC hot line, would have exactly the same effect as hooking across the inductor does. This fact once saved a company I worked for tens of thousands of dollars using the alternate location of the snubber circuit. I would say that makes Thevenizing a pretty powerful tool, wouldn't you?

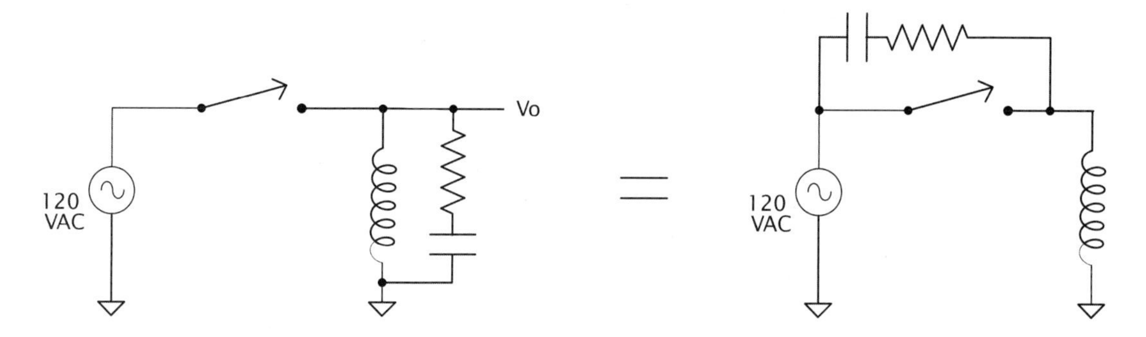

Figure 2-26 *These are equivalent circuits when Thevenized*

Thumb Rules

- The basics are the most important!
- Impedance is resistance at a given frequency.
- V = I * Z.
- Voltage divider rule, Vo = Vi(Rg / (Rg + Rs).
- A capacitor resists a change in voltage but current can change immediately (the inverse of the inductor).
- An inductor resists a change in current, but voltage can change immediately (the inverse of the capacitor).
- A capacitor is to voltage as an inductor is to current.
- Series resistors, series inductors and parallel caps add up.
- Parallel resistors, parallel inductors and series caps use the product over the sums rule.
- When Thevenizing, short voltage sources, open current sources.
- Consider the circuit from the output point of view.
- Insight can be gained by Thevenizing a circuit.

It's About Time

AC/DC and a Dirty Little Secret

AC/DC. It isn't a rock band; it's one of those lovable engineering acronyms. It means alternating current and direct current. These terms came in to being to describe a couple of different modes of electricity. A firm understanding of these two modes will help you in all aspects of engineering. Before we move on to these two modes, we need to establish an understanding of something called *conventional flow* and *electron flow*.

Way back before we even knew that electrons existed, electricity was thought of as a flow of energy. Benjamin Franklin picked a direction for that flow, labeling one side positive and the other negative. (The reason is a whole other story involving wax, wool, and a lot of rubbing.) The presumptions made sense, but it turned out later on, as we came to understand what electrons are, that electricity wasn't really a flow and the electrons actually moved in the other direction.

The truth is the little electrons that produce what we call electricity aren't really a continuous flow of energy; they sort of bump around in these little packets. From an aggregate level, though, these packets of quanta[7] can appear as a flow of energy. It also turned out that these charges moved in the opposite direction of what was previously assumed.

By the time all this was figured out, the conventional flow positive to negative nomenclature was pretty well established. Since all the basic equations work either way, no one has bothered to change this idea. Instead another term is used, electron flow, to describe the way electrons actually move in a circuit.

This seemed like a dirty little secret to me when I first found it out, and I've often wondered if we haven't missed an important discovery along the way due to thinking of electricity in the manner that we do. Considering the world to be flat doesn't cause significant errors with geometry till you are trying to fly a plane to China and you discover what you thought was a straight line is really a curve. So as long as we keep things in perspective, the accepted jargon will do fine.[8] For our discussion, we

[7] Ahh, quantum mechanics, an interesting and whole other topic that we will have to reserve for later.

[8] Don't let that stop you from wondering though; maybe you will discover something new!

will use the conventional flow terminology. We will also consider the effects from an aggregate level, preserving the idea of a flow of current.

Now that you know the dirty little secret of electron flow, let's talk about current and voltage and where it comes from.

Constant Voltage Sources vs. Constant Current Sources

Devices that cause electrons to move are called *sources* since they are the source of electron flow. There are two typical types of sources: voltage sources and current sources. Remember two important things when dealing with these sources.

When dealing with a voltage source, the output will try to maintain the voltage across the load. That is, the voltage at the source will be constant. This means in terms of Ohm's Law that V remains the same at the source. I and R can change, but in the end it must always equal V in terms of Ohm's Law. $V = IR$.

When dealing with a current source, the output will try to maintain the current through the load. That is the current from the source will be constant. These are less common but they do exist and can be used in many situations. Current from the source will remain constant, allowing V to change as R varies, still following Ohm's Law as obediently as any other circuit. $I = V/R$.

The world of electronics is very voltage-centric so you will see voltage sources much more often than current sources. This being the case, I will tend to concentrate more on these types of sources.

Sources can come in two different types: AC or DC. Let's take a closer look.

Direct Current

The term *direct current* is used to describe current that flows in only one direction. I think this makes direct current the simplest to understand, so we should start there.

Direct current moves only one way, from positive to negative.[9] A battery is a common direct-current device. Hooked up to a load such as a resistor, the current will go something like this.

---

[9] Conventional flow considered here.

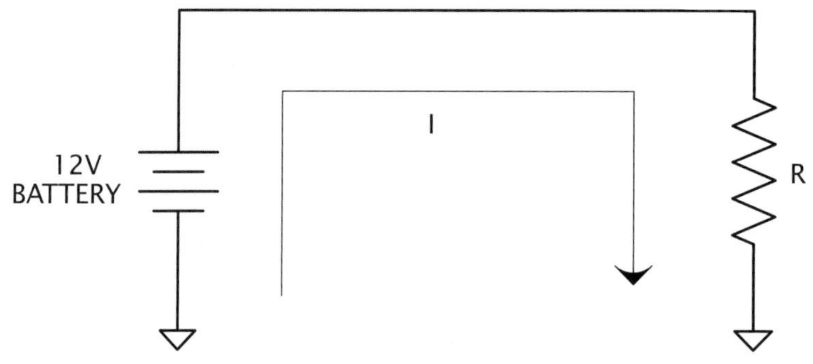

Figure 2-27 *DC current and voltage from a battery*

A battery is also a constant voltage device, so it will apply whatever current is needed to maintain its output voltage. So, we have 12V hooked up to 1Ω of resistance—hey, we just learned how to figure out current on a circuit like that! (More scribbling on a napkin...) That would be 12 amps of current.

A DC source will always try to move current in the same direction. One thing to note is that the current coming out of the source always needs to get back to the source somehow. The ground connection on the schematic should be thought of as a label that connects the signal back to the source. If the signal does not get back to the source, then there is no current flow.[10]

Alternating Current

AC or alternating current came about as the interaction of magnets and electricity were discovered. In an AC circuit, the current repetitively changes direction every so often. That means current increases in flow to a peak, then decreases to zero current flow, then increases in flow in the opposite direction to a peak, then back to zero and the whole process repeats. The current alternates the direction of flow in a sinusoidal fashion, so of course it is called *alternating current*. This type of current most commonly comes from big AC generators at your local hydroelectric dam.

[10] There are those that would argue this point. If you want to know more, do a search for free energy on the internet, but beware—much of it is complete bunk. That doesn't make it a bad read though; it can be humorous and quite thought provoking.

AC power came into being due to this ease of generation. When you move a coil of wire past a magnet, the current first climbs as the strength of the field increases, then as field decreases and switches polarity, the current also decreases and switches polarity. The voltage and current change in the sinusoidal fashion naturally as the coil passes by the magnets.

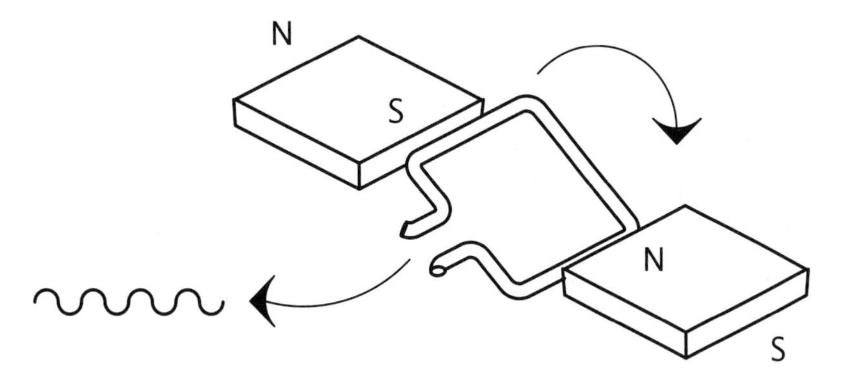

Figure 2-28 *Simple AC generator*

As long as you keep moving the coil, AC power will continue to be generated.

You will see an AC source on a schematic represented by a sine wave squiggle like this:

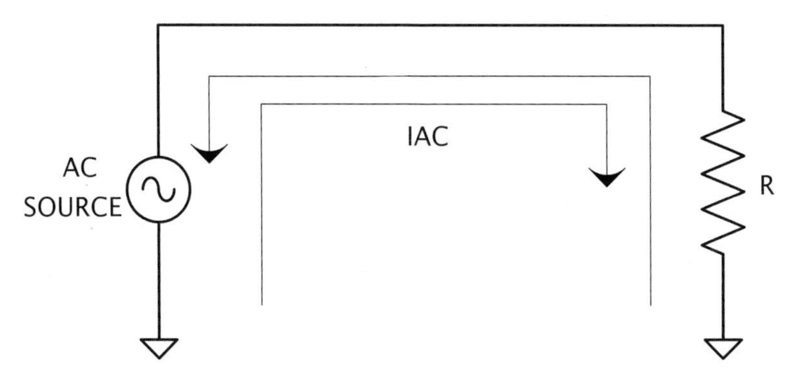

Figure 2-29 *AC voltage and current source*

An interesting side note is that there was some argument when plans were being drawn up to distribute electricity across the nation. Edison (yeah, the famous light bulb guy) wanted to put small DC generators in everybody's home. Another lesser-known genius by the name of Tesla was pitching for AC distribution by wires from a central location. AC made some sense as the voltage could be easily transformed (yeah, you guessed it, with a transformer) from one level to another. That made it possible to jack the voltage up so high that the resistance of the distribution wires had little loss over long distances. There was much debate over the best set-up. One thing that tipped the scales in the direction we have today was the invention of the AC motor by Tesla. Till then, only DC motors had been developed and, since this was before the diode, it wasn't so easy to make AC into DC, so being able to run a motor was a big deal. Although not as famous as Edison, Tesla has left a huge legacy with AC power distributions and AC motors. Just look around your house and count up the AC motors in use. (Of course, there are a few light bulbs around too.)

Back to Capacitors and Inductors Again

What was the rule of thumb for a capacitor? Capacitors impede a change in voltage. Do you remember the rule for inductors? Inductors impede a change in current. The flip side of these rules is that the cap will let current change all it wants, and the inductor will let voltage change all it wants. One fact you can't ignore when it comes to AC sources is that current and voltage are always changing. How fast they change is a function of a term known as frequency. Frequency is the number of cycles of change per second; it has a unit called *hertz*. The higher the frequency is, the faster the change in voltage and current. Now extrapolate for a moment what might happen to a cap in an AC circuit. It follows that a cap will block currents that have zero frequency (like a DC battery) and pass currents that change. The opposite applies to an inductor.

I like to think of it this way: the cap is an infinite resistor at DC or zero frequency. As frequency increases, the "resistance" (technically known as *reactance*) of the cap gets lower and lower, approaching zero. This capacitive reactance is known as X_C and is described by this equation. The unit is ohms, just like a resistor.

$$X_C = \frac{1}{2\pi * f * C}$$ Eq. 2-14

The inductor is just the opposite. It starts with 0Ω of resistance at a zero frequency and then increases to infinity along with the frequency. Inductive reactance is called X_L and follows this equation:

$$X_L = 2\pi * f * L$$ Eq. 2-15

Let's hook them up to an AC source and vary the frequency to see how current flow is affected. This is easy enough to do in a spreadsheet. Just plug the reactance equation into Ohm's Law.

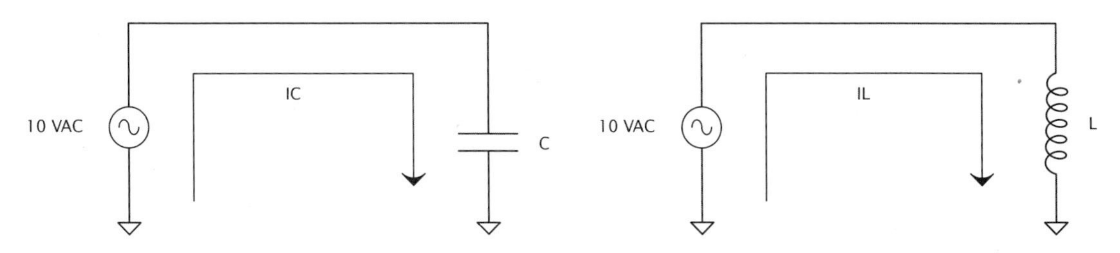

Figure 2-30 *AC source hooked up to cap and inductor*

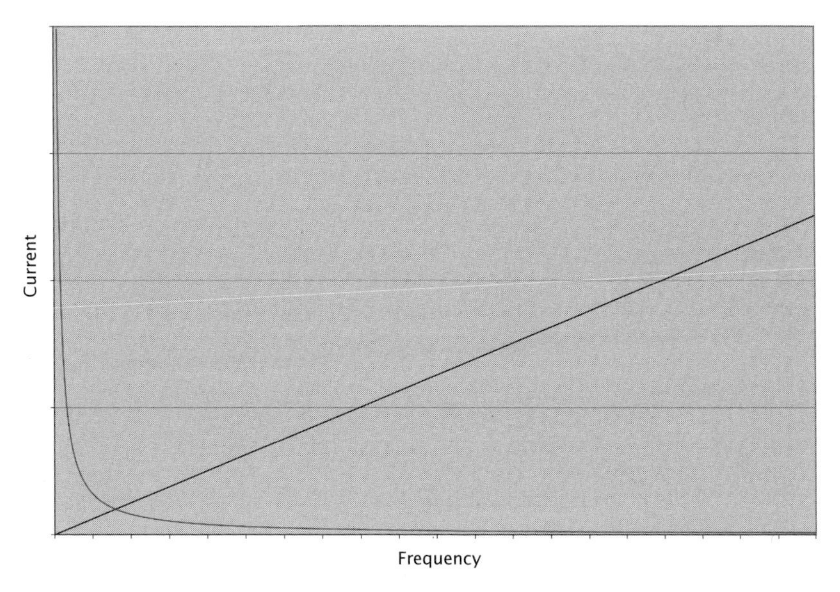

Figure 2-31 *Graph of current over frequency for a cap and an inductor*

So to repeat, the higher the frequency, the easier the current will pass through a capacitor and the harder it becomes for it to get through an inductor.

You might be thinking, "What about that step input we put into the RC circuit before? Isn't it AC? Actually, as weird as it may sound, it is. A really smart guy by the name of Fourier figured out some time ago that hidden in fast-changing signals are all sorts of high frequencies. He proved that the more abrupt the change in the signal, the more high frequencies are present. An in-depth study of this is somewhat beyond the scope of this book, so let it suffice to say that the step input in the previous discussion has a sharp square corner, hidden in which are a whole bunch of high frequencies. These can't get through the cap, so the corner is knocked off, so to speak, leaving the characteristic curve that we saw as the transient response of the RC circuit.

Before we move on we should touch on phase shifts. When both voltage and current are in sync, they are in phase. As we have discussed numerous times, inductors impede a change in current but voltage is not affected, so if you graph the relationship between voltage and current, you will see that the change in current is a little out of sync with the change in voltage. It is said to be lagging behind. The capacitor has the opposite effect (as always), the voltage is delayed relative to the current. In this case the change in current leads the change in voltage. The current isn't magically jumping ahead of the voltage, the voltage is getting behind, but from the voltage point of view it looks like the current is changing first.

So capacitors and inductors are basically resistors that vary based on frequency. The cap delays voltage changes, and the inductor delays current changes. They are opposite in the way they react to the frequency of a signal. Caps block lower frequencies while letting higher ones through, whereas inductors pass lower frequencies while blocking higher ones. Let's see what happens when we hook them up to a resistor.

We really shouldn't leave these two components without discussing phase shifts. When both voltage and current are in sync, they are in phase. As we have discussed numerous times, inductors impede a change in current but voltage is not affected, so if you graph the relationship between voltage and current you will see that the change in current is a little out of sync with the change in voltage. It is said to be lagging behind. The capacitor has the opposite effect (as always); the voltage is delayed relative to the current. In this case, the change in current leads the change in voltage. The current isn't magically jumping ahead of the voltage, the voltage is getting behind, but from the voltage point of view it looks like the current is changing first.

Low-Pass Filters

Consider the following circuit:

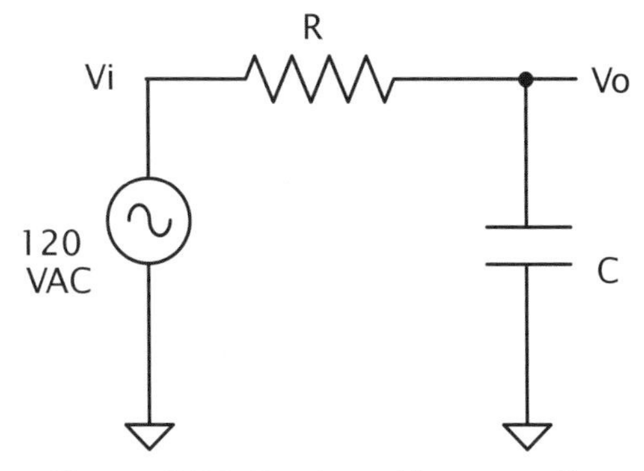

Figure 2-32 *Cap-based low-pass filter*

Note similarities to the RC circuit that we used to first understand the effects of a capacitor. The difference is that now we are going to apply an AC signal to the input rather than the step input we applied before.

This circuit is known as a low-pass filter, and all you really need to know to understand it is the voltage divider rule and how a capacitor reacts to frequency.

If this were a simple voltage divider, you could figure out based on the ratio of the resistors how much voltage would appear at the output. Remember that the cap is a resistor that depends on frequency and try to extrapolate what will happen as frequency sweeps from zero to infinity.

At low frequencies the cap doesn't pass much current, so the signal isn't affected much. As frequency increases, the cap will pass more and more current, shorting the output of the resistor to ground, dividing the output voltage to smaller and smaller levels. There is a magic point where the output is half of the input.[11] It is when the

[11] This is also known as the –3 db down point. I am avoiding decibels for now to limit the knowledge that needs to be assimilated.

frequency equals 1/RC. You might have noticed that this is the inverse of the time constant that we used earlier when we first looked at caps. Kinda cool when it all comes together, isn't it?

This is known as a low-pass filter because it passes low frequencies while reducing or attenuating high frequencies.

You can make a low-pass filter with an inductor and resistor too. Given that the inductor behaves the opposite of a resistor, can you imagine what that might look like?

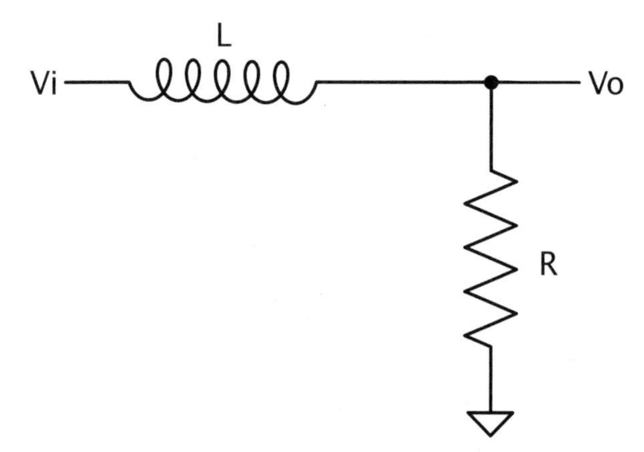

Figure 2-33 *Inductor-based low-pass filter*

That's right; you swap the position of the components. That's because the inductor being the opposite of a cap passes the lower frequencies and blocks the higher frequencies. It performs the same function as the low-pass RC circuit but in a slightly different manner. You still have a voltage-divider circuit, but instead of the resistor to ground changing, the input resistor is changing. At low frequencies the inductor is a short, making the ground resistor of little effect. As frequencies increase, the inductor chokes[12] off the current. Reacting in a way that makes the input element of the voltage divider seem like an increasingly large resistance. This in turn makes the resistor to ground have a much bigger say in the ratio of the voltage-divider circuit.

[12] Without any proof whatsoever, I assert that inductors are sometimes called chokes for the reason that they choke off high frequencies.

To summarize, in the low-pass filter circuits, as the frequencies sweep from low to high, the cap starts out as an open and moves to a short, while the inductor starts out as a short and becomes an open. By positioning these components in opposite locations in the voltage-divider circuit, you create the same filtering effect. The ratio of the voltage divider in both types of filters decreases the output voltage as frequencies increase. All this lets the low frequencies pass and blocks the high frequencies. Now, what do you suspect might happen if we swap the position of the components in these circuits?

High-Pass Filters

Swapping the cap and the resistor in the low-pass circuit creates another type of circuit called a *high-pass filter*. Using your now supreme powers of deduction and intuition, you are thinking to yourself, "I'll bet that means the circuit passes high frequencies while blocking low ones." You are correct and the circuit looks like this:

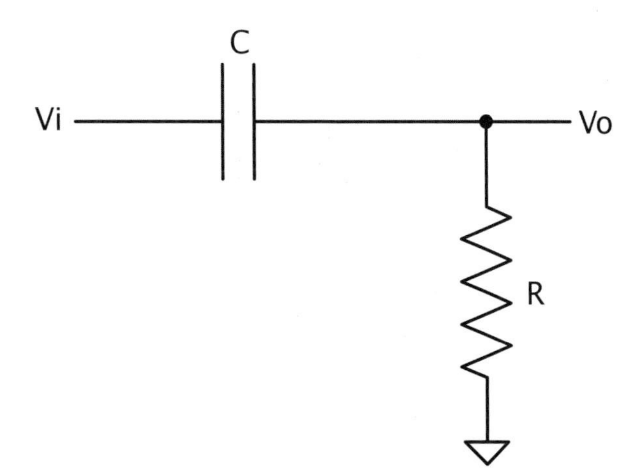

Figure 2-34 *Cap-based high-pass filter*

Hopefully, after the discussion on the low-pass circuit, the operation of this is clear. The cap acts like a larger resistor at low frequencies, making the voltage divider knock down the output. At higher frequencies the cap passes more current as it becomes a short, causing a higher voltage at the output. The inductor version of this circuit looks like this:

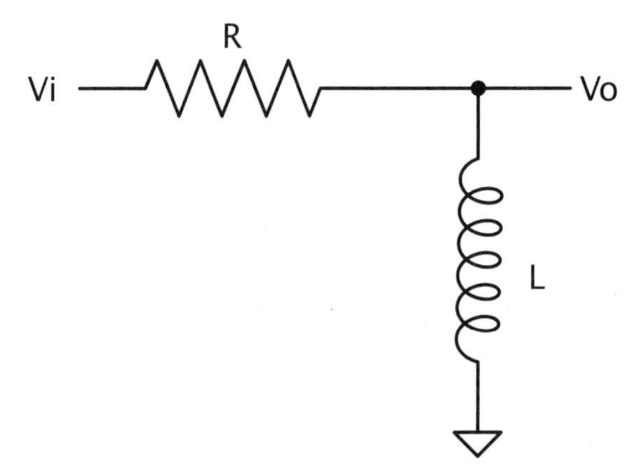

Figure 2-35 *Inductor-based high-pass filter*

As you may have suspected, it is the inverse, circuit-wise, of the RC high-pass filter. Another little bit of serendipity is the fact that the half voltage output point[13] is also at 1/tau, just like the low-pass filters.

To sum it up, the high-pass and low-pass filters take advantage of the frequency response of either a capacitor or an inductor. This is done by combining them with a resistor to create a voltage divider that attenuates the unwanted frequencies while allowing the desired ones to pass. Some cool things happen when we put the two reactive elements together. You can create notch and bandpass filters where a specific band of frequencies is knocked out, or a specific band is passed while all others are blocked. The phenomenon of resonance also occurs in what is called a *tank circuit*, where you have a capacitor combined with an inductor. Just like the spring-mass example in Chapter 2 the tank circuit will oscillate current and back and forth from one component to the other.

Active Filters

So far we have been studying passive filters. A passive component is one that is not powered externally. Being passive, they are subject to an effect known as loading.

[13] This point is also known as the roll-off or 3 db down point.

This means anything you hook up to the output can affect the performance of the filter. Take a low-pass RC filter, for example, and hook a resistor up to it, like this:

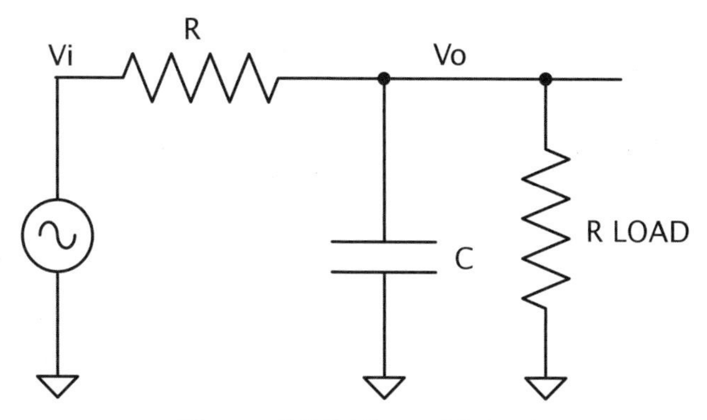

Figure 2-36 *Filter with load*

This resistor on the output is a load. It could be another part of the circuit or any number of things, but the point is that it acts like a resistor to ground. How does this affect the RC filter performance? To understand that, let's Thevenize it. We start by shorting the voltage source to ground. This is done with AC sources the same as DC, so the circuit would look like this:

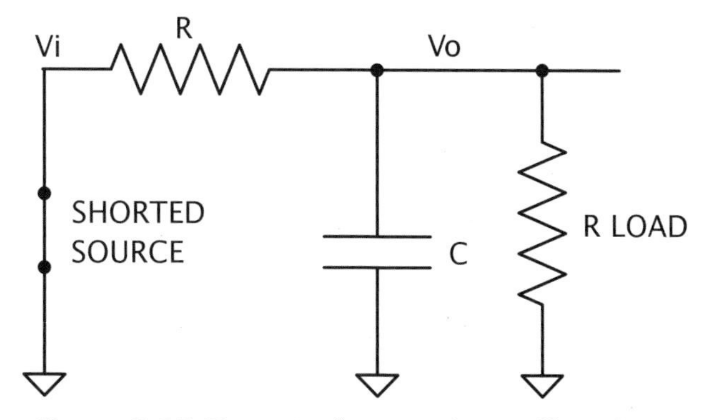

Figure 2-37 *Thevenized circuit shows effect of load*

As I like my examples to use real numbers, let's make up some values. Let $R = 10K$, let $Rload = 10K$ and $C = 0.1$ µf.

When you Thevenize a circuit you reduce all the parts into one where possible. In this case the resistors are in parallel, so apply the parallel rule to the resistors and you get a value of 5K Ω. Did you notice that the *R* value has changed considerably due to the load on the circuit? What may seem counterintuitive at first is the fact that the time constant of this circuit is a function of the Thevenized version that we just derived. So without the load, tau would have been 10K * 0.1 μs or 1 ms.

With the load, it is 0.5 ms, half of what it was before! Since the output of this filter depends on τ, we can see that the load has affected it significantly. A way to avoid this problem is to add an active component to the design, making it into an active filter. In adding such a component, the basic idea is to minimize this loading effect to a point that you get a nice predictable response. The output of the active filter is such that no matter what load you put on it, it does not affect the response of the filter.

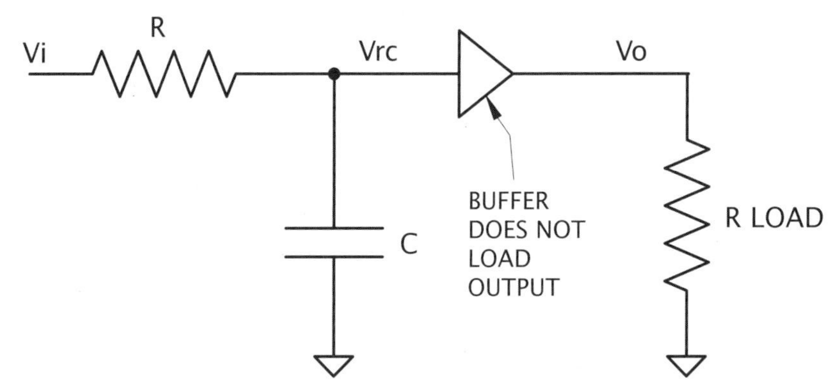

Figure 2-38 *Active buffer eliminates the effect of the load*

The input of this active device has a very high impedance. In this case, that means it is comparable to a 10-meg resistor. Hooking that up to the RC filter will have little effect on the time constant of this circuit as long is it is significantly larger than the *R* value in the circuit. The buffer in this circuit will output a voltage that matches the voltage on the input. It will *buffer* the signal; no matter what you hook up to the output, the filter will not be affected. This is one of the simplest active filters, but the principle with all of them is the same—include an active element to preserve or enhance the integrity of the filter.

Thumb Rules

- Electrons move from negative to positive.
- Direct current flows in one direction; it has a zero frequency component.
- Alternating current changes direction of flow repetitively.
- Direct current has a frequency component of zero
- Inductors and capacitors can make both low-pass and high-pass filters when combined with a resistor.
- Inductors and caps in the same circuit will oscillate
- Active filters add components to preserve or enhance the integrity of the filter.

Beam Me Up?

Electrons are everywhere, or maybe it is better said that the effects of electrons are seen everywhere. There are invisible fields of force all around us that are caused by those pesky little devils. These fields warrant discussion as they are a factor in how we deal with electricity. They can store energy and affect the world around them in various ways, so it is good to build an intimate knowledge of these fields and how they interact.

No, we can't beam people[14] on and off the planet like they do on Star Trek, but there are some invisible fields out there that create the effects that we see in electric circuits. Understanding these fields will only help develop the intuitive skills we are working on.

The Magnetic Field

This is the most well-known of the two fields that we are going to discuss. Who hasn't experienced the force of a magnet, sticking a note to the fridge or feeling the power of two repelling magnets? Back in the 1820s, a man by the name of Hans Oersted noticed his compass read funny every time he switched on a current in a wire. Eventually it was figured out that a moving electron (such as the current in a wire) creates a magnetic field perpendicular to the direction of electron movement.

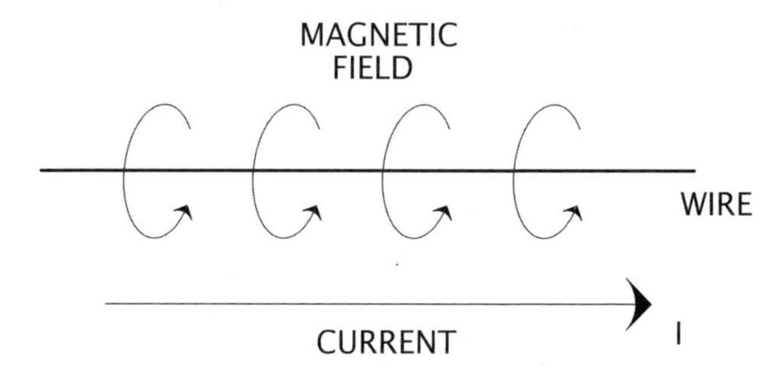

Figure 2-39 *Magnetic field caused by current in a wire*

[14] Note I said people; physicists have "beamed" a quantum bit from one point to another. Some say teleporting an atom could happen in just a few years.

This field is identical to the field surrounding a permanent magnet. In fact, if you coil the wire like this, the magnetic field lines align and reinforce each other, making it even more like a permanent magnet.

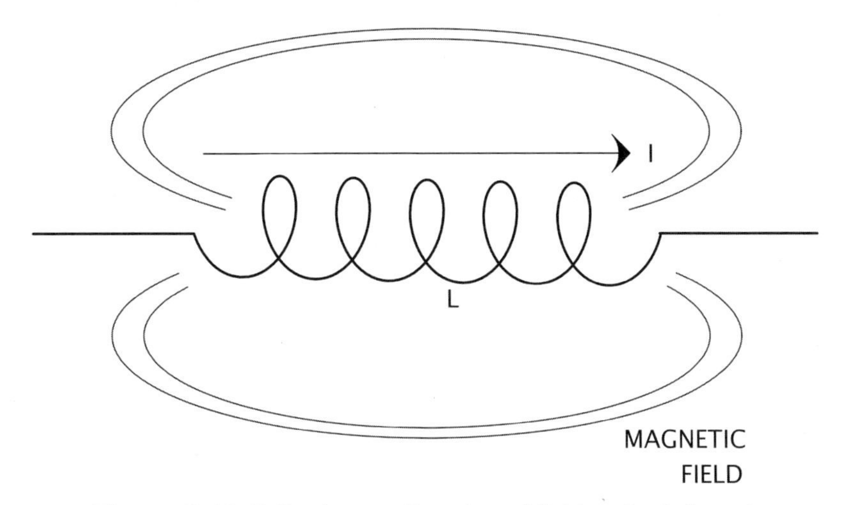

Figure 2-40 *Coils change direction of field and reinforce it*

Electromagnets, as they are called, are pretty cool since they can be switched on and off, unlike permanent magnets. Another important fact is that not only does a current moving through a wire create a magnetic field, but the opposite is also true. A changing magnetic field can create or induce a current in a wire. A coil of wire is known as an inductor for this reason. Energy is stored in an inductor as a magnetic field. It is like a rubber band that is stretched as you apply current. When the current is shut off it snaps back, and energy is given up as the magnetic field collapses (it is changing as it goes away). This collapse induces a current in the wire. Consider this circuit:

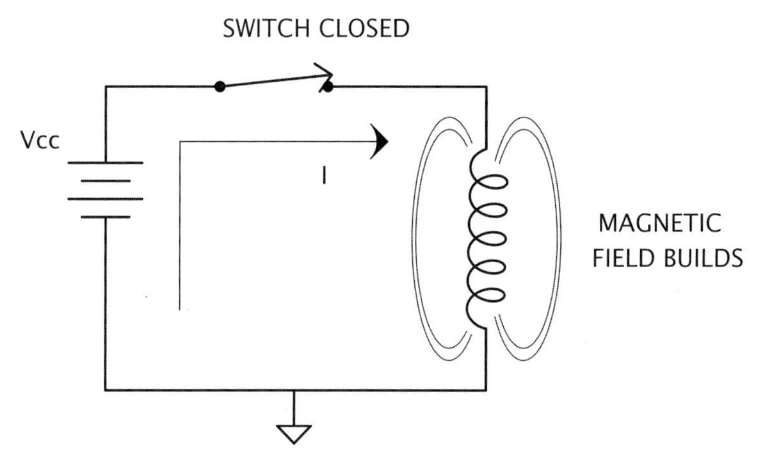

Figure 2-41 *Building magnetic field resists current change*

When the switch is closed, current flows and a magnetic field is created. It is the creation of the magnetic field that causes the inductor to "resist" the change in current, as we learned it does earlier. The flip side of that also happens. If we open the switch, the change in the field as it collapses would like to keep the current flowing in the inductor.

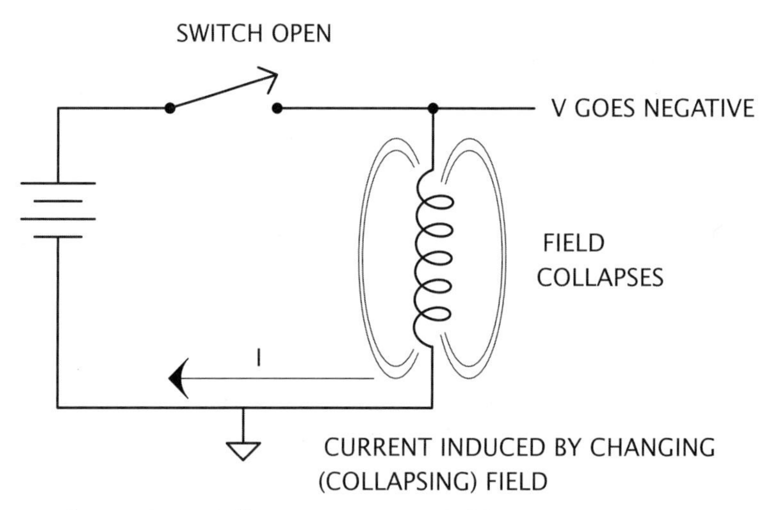

Figure 2-42 *Collapsing magnetic field generates a current*

If there is no place for this current to go, the voltage across the inductor will increase instantaneously and then dissipate as the induced current drops off with the drop of the magnetic field. Take a look at this graph of the current and voltage changing in this inductor circuit as the switch opens and closes:

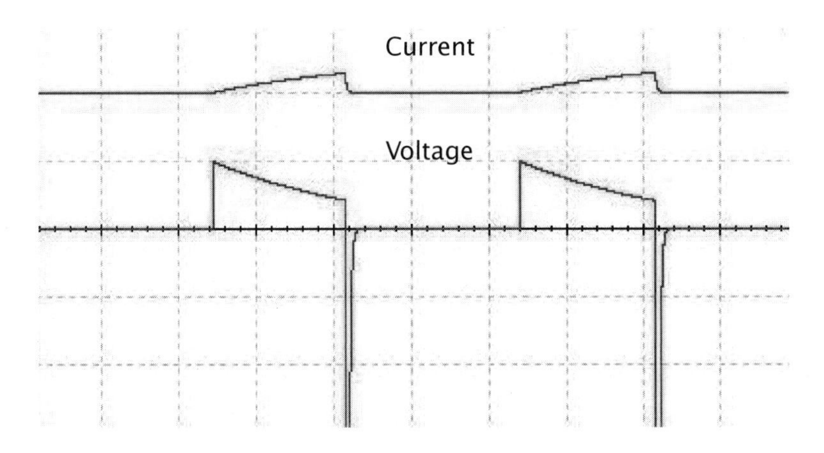

Figure 2-43 *Voltage and current changes as an inductor is switched in and out of circuit*

Induction is also the fundamental principle that a transformer uses. The magnetic field, as it is created on one side of the transformer, induces a current on the other side of the transformer. When the field reduces and switches direction, a corresponding current is induced at the output.

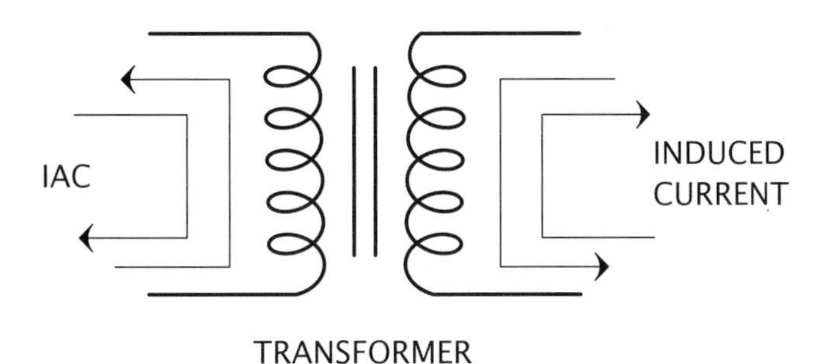

Figure 2-44 *A transformer uses changing current on the input to induce current on the output*

The ratio of turns on each side of the transformer controls the ratio of voltage from input to output. A 10:1 ratio will take 120V on one side and create 12V on the other. Note also that while voltage goes down, current goes up, making a transformer kind of like a gear train or lever in the mechanical world. Power into it is the same as power out of it (minus losses of course). Voltage times current in equals voltage times current out. This is akin to the rule of force times distance on one side of a lever equals force times distance on the other side.

One side of a transformer is an inductor. An inductor is simply a coil of wire as we learned earlier. The number of turns of wire controls the concentration of the magnetic field. The core of the inductor also has the effect of concentrating the field. The material in the core can become saturated, meaning it cannot concentrate the field any more tightly than it has.

The important things to remember are that current creates a magnetic field and a changing magnetic field creates a current. The changing field can be externally applied from a moving magnet, the input side of a transformer, or from the collapse of the field just created by the current. *Current* and *magnetic* fields are closely connected.

The Electric Field

Also called the *electrostatic field*, the electric field is not as commonly known as the magnetic field. In the same way that current is connected to the magnetic field, voltage is connected to the electric field. That leads to a good rule of thumb to remember: *current is magnetic* and *voltage is electric*.

The electric field comes from electric charges, both positive and negative. In a way analogous to like poles on magnets repel and opposite poles attract, like charges repel and opposite charges attract. Any molecule or atom can be neutral (no net charge), positively charged or negatively charged. The accumulation of these charges is what is known as voltage. One way to think of it is that the *charges* are the *voltage* making the electric field, and *movement of those charges* is called *current* and creates the magnetic field.

As an inductor is a way of concentrating a magnetic field, a capacitor is a way of concentrating an electric field. Capacitors are made by two collectors or plates separated by a material that will not conduct electricity, also known as a dielectric. The symbol of a capacitor mimics this construction:

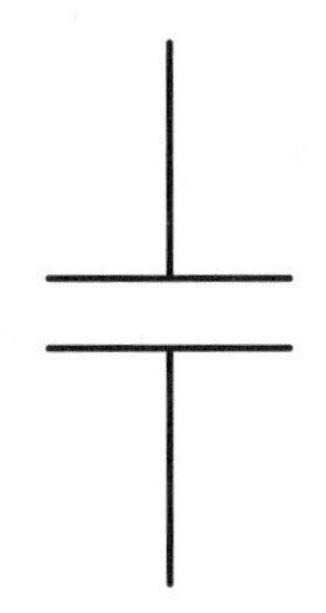

Figure 2-45 *Capacitor symbol*

Because of the dielectric, current cannot flow across the capacitor, so all the charges build up on one side of the cap, kind of like a fifty-car pile-up on the freeway.

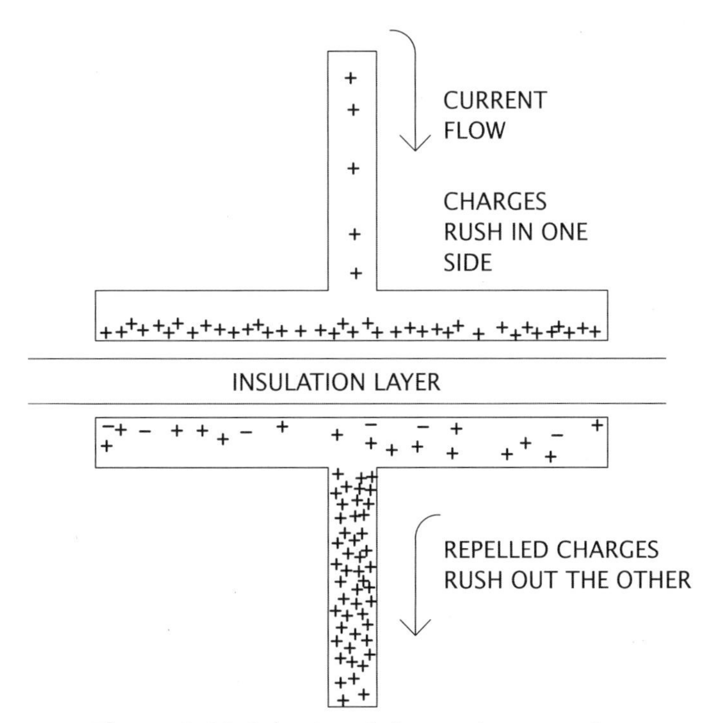

Figure 2-46 *Behavior of charges in a capacitor*

As the charges pile up on one side, the electric field builds up, causing all the like charges on the other side of the cap to go rushing away. Once it all comes to rest, there is an equal amount of opposite charges on the other side of the cap. In this way the capacitor stores a charge of voltage on the plates of the capacitor.

How much charge a cap can store in an electric field is a function of the area of the plates. The amount of voltage it can store is dependent on the strength of the dielectric. If you exceed the capability of the insulation, the dielectric will break down and a charge will cross the gap. The same thing happens on a stormy day. During a thunderstorm charges build up in the clouds and the ground in the same way they do on either side of a capacitor. A lightning strike is a large-scale version of what happens when the insulation or dielectric in a capacitor breaks down.

In the same way current creates a magnetic field, voltage creates an electric field. As the magnetic field can store energy, the electric field can also store energy. As the magnetic field dissipates, it tries to maintain current. As the electric field dissipates, it tries to maintain voltage. Voltage and electric fields are closely connected.

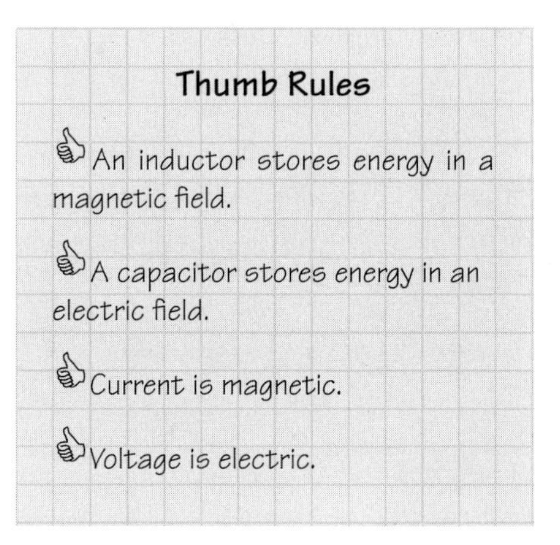

Keep It Under Control

Control theory is typically left till later in most education programs, as it is considered a more advanced topic. Control systems, however, are a very common application in the electronic realm. Think about it; I'll bet most of the time you design a device to control something you don't want to think about, something that automatically does what it should without intervention. On top of that, control theory turns out to be useful in more than just clear-cut control applications—understanding op-amps and designing power supplies, for example. As a basic knowledge of this concept will help in many aspects, it seemed prudent to dedicate a few pages to it.

The System Concept

A system is anything with an input and an output. The idea is simple—take the input, shake it, squeeze it, do whatever and then send it to the output. It can be represented with a block diagram, something like this:

Figure 2-47 *The magic box inside a system*

All the magic happens inside the box. This magic is called the *transfer function*. The transfer function is equal to the input over the output. It is the equation that you process the input through to get the output, so the following is true:

$$Output = Magic * Input \qquad \text{Eq. 2-16}$$

A little algebra yields:

$$\frac{Output}{Input} = Magic \qquad \text{Eq. 2-17}$$

Now you know how to find what the magic inside the box is, and sometimes it is just that easy. Let's try a simple example to see how. You put 12 miles into the input, wait...chugga, chugga, ding!...and out pops 19.32 km. As you might have guessed, the magic in this box is a metric converter, but what is the transfer function? According to the equation above we just divide the output by the input. That would be:

$$\frac{19.32\ km}{12\ miles} = 1.61\ \tfrac{km}{mile}$$

Eq. 2-18

The magic in our converter box looks like this:

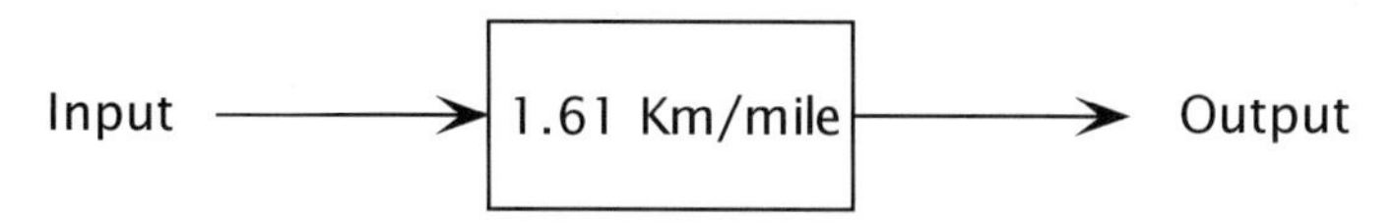

Figure 2-48 *The magic revealed*

Please notice the units made it into the box. This helps identify what type of units will work at the input and what you will get at the output. Hopefully a little voice in your head is saying, "Isn't this a rehash of the unit math chapter?" It is, but this is a more formalized concept with some neat touches such as the cool little boxes you draw to help you understand the system. The next question you should ask is "How does this apply to electronics?" Well, from the most basic to the most complex, you can represent any circuit with one of these magic (ok, some texts call them black) boxes. We'd better do another example. Take a resistor. A resistor can be thought of as a current-to-voltage converter. Put current into the input, apply magic, get voltage at the output. What would be the transfer function of that? If you mumbled a phrase with the words "Ohm's Law" anywhere in it, you are probably right.

$$R = \frac{V}{I}$$

Eq. 2-19

That would make the block diagram of the resistor look something like this:

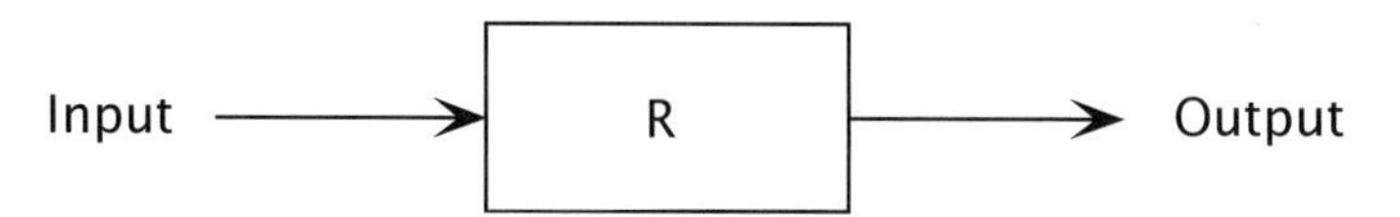

Figure 2-49 *System diagram of a resistor*

In this transfer function R is the value of the resistor in ohms, just in case you didn't guess. (Note that the 1 unit of ohms equals 1 unit of volts divided by 1 unit of amps, like good old Ohm's Law says it does.)

Let's step the idea up to something a little more complex, like a voltage divider. We already know the equation for this. It is:

$$Vo = Vi\frac{Rg}{Rg + Ri} \qquad \text{Eq. 2-20}$$

Do you remember what *Vo* stands for? How about *Vi*? They are voltage output and voltage input. So let's just use a little algebra to figure this out. The transfer function is equal to the output over the input, like this:

$$\frac{Vo}{Vi} = \frac{Rg}{Rg + Ri} \qquad \text{Eq. 2-21}$$

The block diagram would look like this:

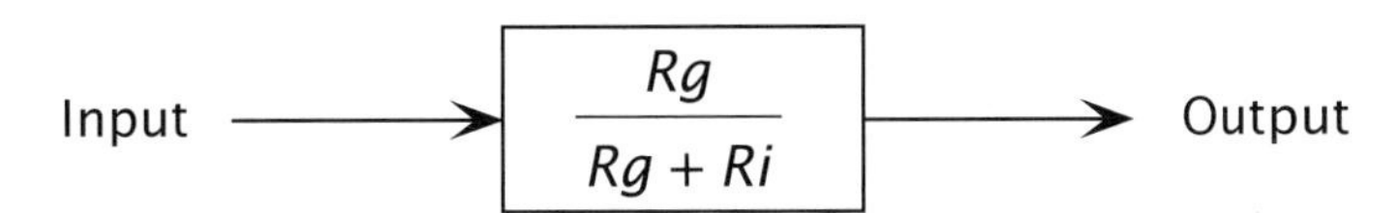

Figure 2-50 *System diagram of a voltage divider*

This same concept can be used to describe all of the circuits that we have seen so far. You may see block diagrams of this type where C or *L* has a little "s" by it. This is a mathematical trick known as a Laplace transform. It is used to simplify problem solving. If you transform all that time constant and frequency stuff using Laplace pairs, you can treat the transformed equations with simple algebra, and then transform it all back when you are done. Laplace transforms are beyond the scope of this text, but do take note that the "s" rolls up all the frequency response of capacitors and inductors into a domain that can be handled easily when the equations get complex.

We can describe any system as a magic box with an input and an output. One way to determine what is in the magic box is to put a known signal into the input. Let's take a look at what is probably the most important stimulus you can apply to the input of the magic box.

The Step Input

The idea behind the step input is to understand the output of a system by its response to a given input. The step input is an instantaneous change of state from a value of zero to another predetermined value. It looks like this:

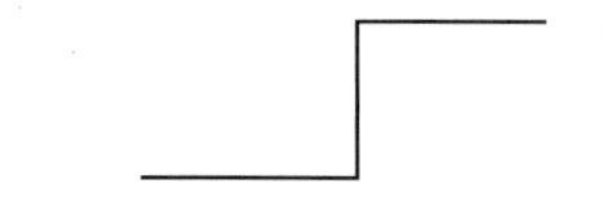

Figure 2-51 *The step input*

The output will change in some manner predictable (one hopes!) by an equation. This is known as the response. The better you know the response of various components to this step input, the better you will be able to apply intuitive signal analysis.

Let's go over the RC circuit again. The equation we learned earlier is:

$$Vo = Vi\left(1 - e^{-\frac{t}{rc}}\right)$$

Eq. 2-22

To turn that into the transfer function for the magic box, all we have to do is move *Vi* to the other side of the equal sign like this:

$$\frac{Vo}{Vi} = \left(1 - e^{-\frac{t}{rc}}\right)$$

Eq. 2-23

We now plug that into the box. When we put the step input into one side we get the familiar RC curve on the output.

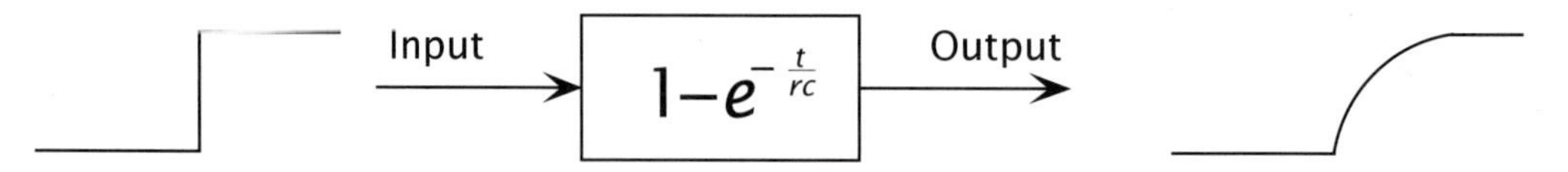

Figure 2-52 *Step input into magic box with RC circuit inside*

While it is nice to see the same conclusion from a different approach, where the system concept comes into its own is when we create a feedback loop.

Feedback

One of the neatest applications of control theory occurs when we implement feedback. Feedback is the process of using the output of the "magic box" as some portion of the input. Feedback comes in two flavors: positive and negative. They can be thought of in terms of interaction.

Positive Feedback

Positive feedback encourages or reinforces a behavior while negative feedback corrects or controls a behavior. For example, if your son is doing a good job in a soccer game and you cheer him on, which encourages him to try even harder, this is positive feedback. Positive feedback reinforces the behavior you desire. In the case above it will encourage him to try as hard as possible. In fact, in a perfect world he will keep trying until he is giving it all he can. The same thing happens with positive feedback in control theory. Output is fed back to the positive input. This has the effect of increasing the output which is fed back to the input, which will increase the output and so on (reinforcing the behavior) until the output is as high as it can go.

Since positive feedback reinforces the signal, the output can often "stick" at the rail. For this reason the amount of positive feedback allowed is typically very limited, allowing only small changes to the input. These small changes can create a feature called *hysteresis*.

Another interesting thing that can happen due to this reinforcing behavior occurs when delays are created in the positive feedback loop. Think about it for a moment—what will happen if this signal is delayed a bit? If you time it right, the signal to change the output can be made to occur at the input when the output is already moving in the opposite direction. When this happens you have created an oscillator.

Now while positive feedback is great for controlling toddlers, when it comes to circuits, if you want to control something you need negative feedback.

Negative Feedback

Negative feedback is a control situation. Let's go back to the soccer analogy for a moment. In this case, your son kicks the ball too far ahead of the player he is passing it to. You tell him to shorten up his pass. If he is not passing far enough, you tell him to lengthen it out. Based on how close the actual result is to the desired result, a corrective signal is fed back to the input. This corrective signal has a negative impact on the output, hence the term negative feedback.

Humans have an innate ability to handle negative feedback.[15] You probably experienced it this morning as you drove your car to work. If you drifted too close to the edge of the lane on the freeway, you processed a little negative feedback, resulting in a corrective signal to bring the car back to the center of the road. If you didn't, you are probably reading this in the passenger seat of the tow truck as your mangled car is hauled home!

Negative feedback is often used to create controlled amplifiers and filters.

We will get into some details of negative and positive feedback and how they work using op-amps a bit later in the book.

Open-Loop Gain vs. Closed-Loop Gain

When you cut the feedback signals out of the "loop" the gain of the system is known as the open-loop gain. This is to distinguish it from the closed-loop gain, or the gain of the system when the feedback is in place.

High open-loop gains in conjunction with negative feedback will minimize errors in amplifier and filter circuits.

[15] Unless of course it is coming from your significant other; then we tend to have serious system-wide erratic behavior.

Thumb Rules

- Lump everything into one magic box.
- The gain or magic equals the output over the input.
- Feedback loops can be added to these systems to create different results.
- Positive feedback tends to latch up or go to an output rail.
- Positive feedback delays can create oscillations.
- Negative feedback signals are corrective in nature.
- Negative feedback creates a controlled output.
- The gain of the system from input to output when the feedback is disconnected is known as the open-loop gain.
- The gain of the system when feedback is in place is known as the closed-loop gain.

CHAPTER 3

Pieces Parts

It takes parts to make a circuit, and lots of pieces too. The better you know how these "pieces parts" work, as my farmer friend likes to say, the better stuff you will build.

Partially Conducting Electricity

Semiconductors

Texts are available that can give you the quantum mechanical principles on which a semiconductor works. However, in this context I think the better thing to do is to give you a basic intuitive understanding of semiconductor components.

First, what is a semiconductor? Conductor in this case refers to the conduction of electricity. Think of a semiconductor as a material that partially conducts electricity, or a material that is only semi-good at conducting electricity. It is similar to the resistor that we just learned about; it's a component that will conduct electricity but not easily. In fact, the more you push through it, the hotter it gets as it resists this flow of electricity.

Before we move on, there is one other point to make. The world of semiconductor devices can be grouped into two categories: current driven and voltage driven. Current-driven parts require current flow to get them to act. Voltage-driven devices respond to a change in voltage at the input. How much current or voltage is needed depends on the device you are dealing with.

Diodes

We will start with the diode. A diode is made of two types of semiconductors pushed together. They are known as type P and type N. They are created by a process called *doping*. In doping the silicon, an impurity is created in the crystal which affects the structure of the crystal. The type of impurity created can cause some very cool effects in silicon as it relates to electron flow. Some dopants will create a type N structure in which there are some extra electrons just hanging out with nowhere to go. Other dopants will create a type P structure in which there are missing electrons, also called *holes*. So we have one type, N, that will conduct electricity with a little effort. We have another type that not only does not conduct, but actually has holes that need filling. A cool thing happens when we smash these two types together; a sort of one-way electron valve is created known as the *diode*.

Figure 3-1 *The PN junction of the diode*

Due to the interaction of the holes and the free electrons, a diode allows current to flow in only one direction. A perfect diode would conduct electricity in one direction without any effect on the signal. In actuality, a diode has two important characteristics to consider: the forward voltage drop, and the reverse breakdown voltage. (See Figure 3-2.)

Forward voltage

The forward voltage is the amount of voltage needed to get current to flow across the diode. This is important to know because if you are trying to get a signal through a diode that is less than the forward voltage, you will be disappointed. Another often overlooked fact is the forward voltage times the current through the diode is the amount of power being dissipated at the diode junction. If this power exceeds the wattage rating of the diode, you will soon see the magic smoke come out and the diode will be toast.

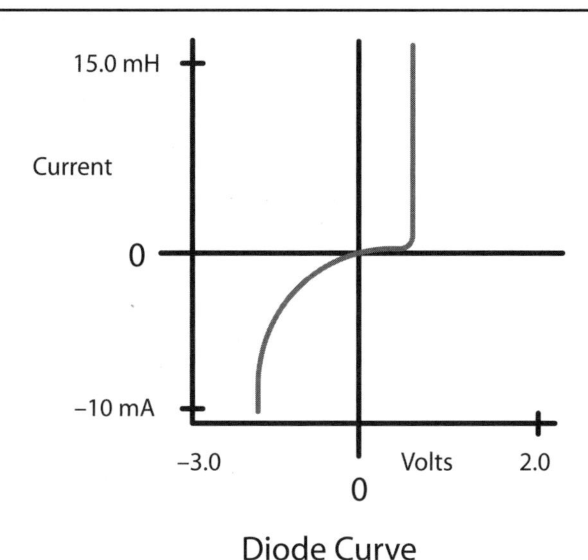

Figure 3-2 *Typical diode voltage/current response*

For example, you have a diode with a forward voltage rating of 0.7V and the circuit draws 2A. This diode will be dissipating 1.4W of energy as heat (just like a resistor). Verifying that your selection of diode can handle the power needed is an important rule of thumb.

Reverse Breakdown Voltage

While a perfect diode could block any amount of voltage, the fact is, just like humans, every diode has its price. If the voltage in the reverse direction gets high enough, current will flow. The point at which this happens is called the *breakdown voltage* or the *peak inverse voltage*.[1] This voltage usually is pretty high, but keep in mind it can be reached, especially if you are switching an inductor or motor in your circuit.

[1] It is interesting to note that there is a type of diode called a zener in which this breakdown voltage is controlled and counted on. I would further stress the importance of calculating power in a zener. In this case, however, it is the zener voltage or the reverse voltage that you must multiply by current to calculate the power dissipation. Isn't zener a cool word to say?!

Transistors

The next type of semiconductor is made by tacking on another type P or type N junction to the diode structure. It is called a *BJT*, for bipolar junction transistor, or *transistor* for short. They come in two flavors: NPN and PNP. I presume you can guess where those labels came from.

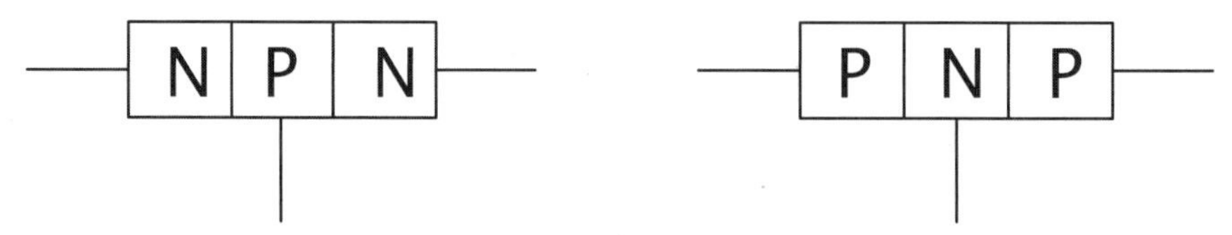

Figure 3-3 *Smash diodes together to make a transistor*

At first glance you would probably say, "Isn't this just a couple of diodes hooked up back to back? Wouldn't that prevent current from flowing in either direction?" Well, you would be correct. It is a couple of diodes tied together and, yes, that prevents current flow. That is, unless you apply a current to the middle part, also known as the base of the transistor. When a current is applied to the base, the junction is energized and current flows through the transistor. The other connections on the transistor are called the *collector* and the *emitter*. The NPN needs current to be pushed into the base to turn the transistor on, while the PNP needs current to be pulled out of the base to turn it on.[2] In other words the NPN needs the base to be more positive than the emitter, while the PNP needs the base to be more negative than the emitter. Remember the similarity to the diode? It is so close that the base-to-emitter junction behaves exactly like a diode, which means you need to overcome the forward voltage drop to get it to conduct.

[2] In this case I am referring to conventional flow, as it is called. That is why I used the word current. Back when electricity was discovered, with the instruments they had it appeared to be a flow of energy. A positive connection and a negative connection were selected and current was postulated to flow from positive to negative. In reality the flow of this current was not really a "flow" and it went the other direction. But the formulas they worked out still solved the problems correctly so no one has changed the nomenclature. If you want to talk about what the electrons actually do, you call it electron flow, and if you are talking about the way all the formulas and diagrams are written you should think in terms of conventional flow.

Whoever is in charge of making up component symbols has made it easy for us. There is a very "diode-like" symbol on the emitter-to-base junction that indicates the presence of this diode. Also, please note that I keep talking about current into and out of the base of the transistors. Transistors are current-driven devices; they require significant current flow to operate. Most times the current flow needed in the base is 50 to 100 times less than the amount flowing through the emitter and collector, but it is significant when compared to what are called *voltage-driven devices*.

Transistors can be used as amplifiers and switches. We should consider both types of applications.

Transistors as Switches

In today's digital world, transistors are often used as switches amplifying the output capability of a microcontroller, for example. Since this is such a common application, we will discuss some design guidelines for using transistors in this manner.

Saturation

When using a transistor as a switch, always consider if you are driving the device into saturation. Saturation is when you are putting enough current into the base to get the transistor to move the maximum amount through the collector. Many times I have seen an engineer scratching his head over a transistor that wasn't working right, only to find there was not enough current going into the base.

Use the Right Transistor for the Job

Use an NPN to switch a ground leg and a PNP to switch a Vcc leg. This may seem odd to you at first. After all they are both like a switch, right? Well, they are like a switch, but the diode drop in the base causes an important difference, especially when you only have 0V to 5V to deal with. Consider the following two designs:

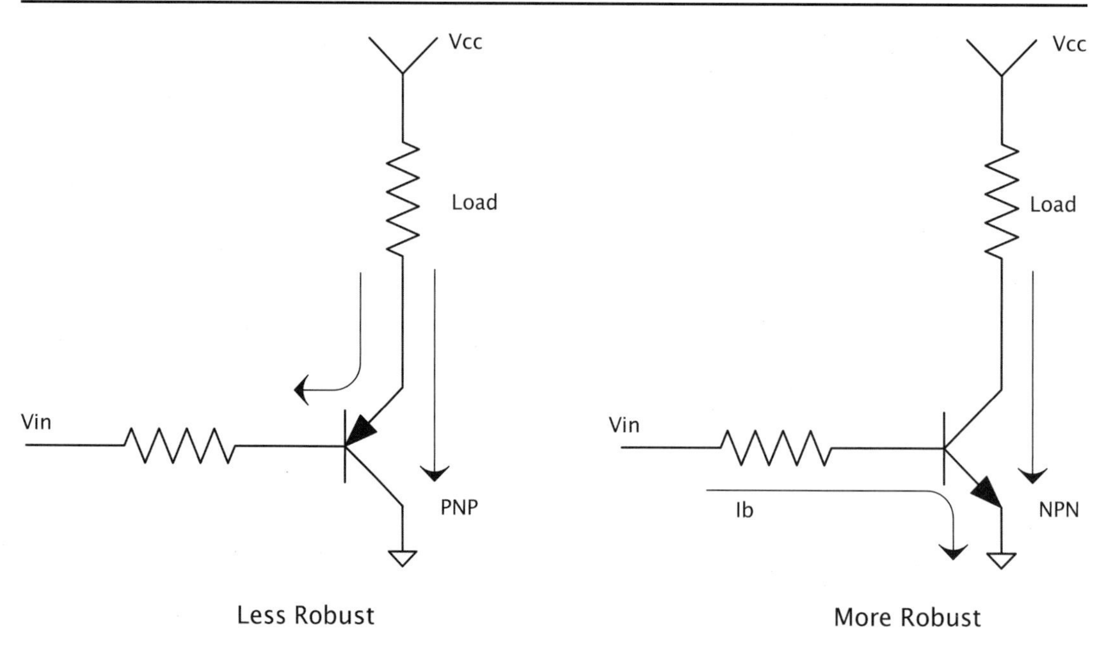

Figure 3-4 *Comparison of different transistors in the same circuit*

Let's do a little ISA[3] on the less robust circuit. As you decrease the voltage at the input, current will flow through the base, but the emitter base junction is a diode, right? That means whatever voltage the base is at, the emitter is always 0.7V higher. Even if you pull the input to 0V exactly, since current has to flow, the voltage at the base will be a little higher. The voltage at the emitter will be 0.7V above that. Notice now that any voltage change at this point will be reflected at the output. Now contrast that with the more robust design. When you pull the signal at the input low, current will flow through the base just like the other design, but do you see the difference? In the second design the input voltage can vary quite a bit, and as long as the transistor is in saturation, the voltage drop at the output from collector to emitter will remain the same.

[3] Intuitive Signal Analysis—see Chapter 3. I have to get an acronym out there if I am to change the engineering world. Too bad all good acronyms mean more than one thing!

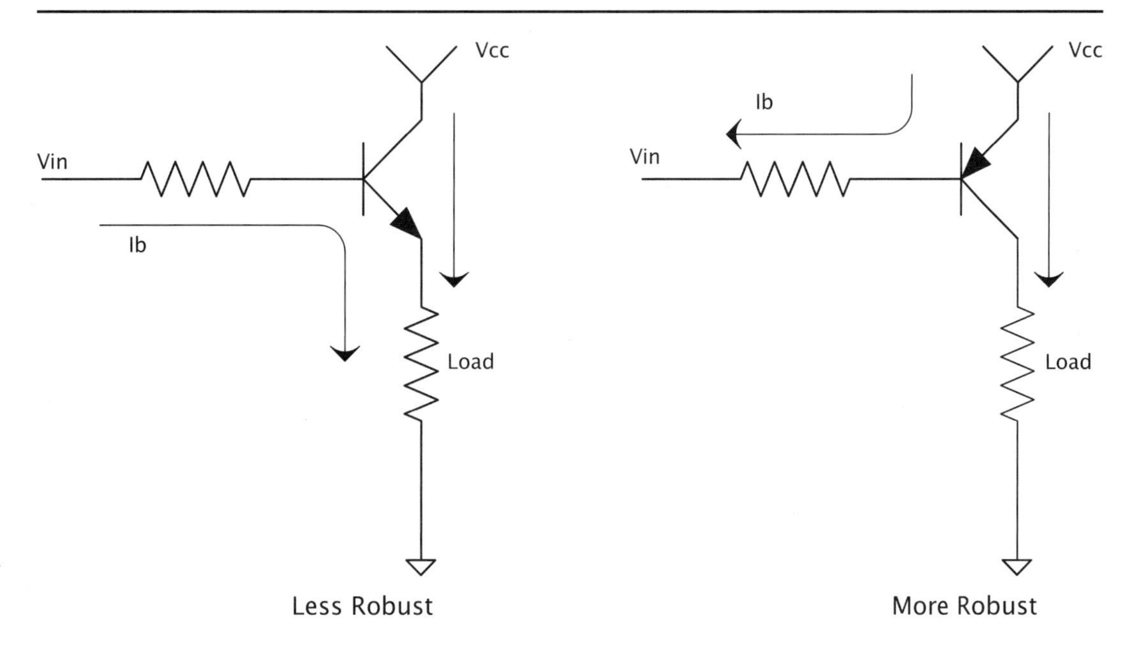

Figure 3-5 *Comparison of different transistors in the same circuit*

The PNP transistor works best in the opposite configuration. For a switching application it is more robust when controlling the Vcc leg of the load. In both cases, turning the transistor off is not too difficult, just get the base with in 0.7V of the emitter and the current will stop flowing.

Transistors as Linear Amplifiers

Transistors can also be used as linear amplifiers. This is because the amount of current flowing through the collector is proportional to the current through the base. This is called the *beta* or *HFE* of the transistor. For example, if you put 5 µA into the base of a transistor with a beta of 100, then you would get 0.5 mA of collector current. Making this work correctly depends on keeping the transistor operating inside a couple of important limits.

One limit is created by the diode in the base-to-emitter connection. This diode needs to remain forward-biased for the transistor to amplify linearly. It is also important to keep the transistor out of saturation. This can push the transistor out of its linear region, creating funny results such as clipping. What all this means is that setting up linear transistor amplifiers can be a bit of a trick. You need to pay attention to biasing and the HFE, which unfortunately varies considerably from part to part. These days I rarely use transistors alone as linear amplifiers for two reasons: the first is the amount of variation from part to part mentioned above (a real issue when you make millions of circuits), and the second is the fact that operational amplifiers (which we will discuss later) are so inexpensive[4] and easy to use. If you need the power capability of a transistor, you should try teaming it up with an op-amp to make life easier!

FETs

FETs or field effect transistors were developed more recently than transistors and diodes. Why come up with something new? Simple—FETs have some properties that makes them very desirable components. The primary reason they are so slick is that the output of a FET is basically a resistance that varies depending on the voltage at the input. The outputs on a FET are called the *drain and source*, while the input is known as the *gate*.

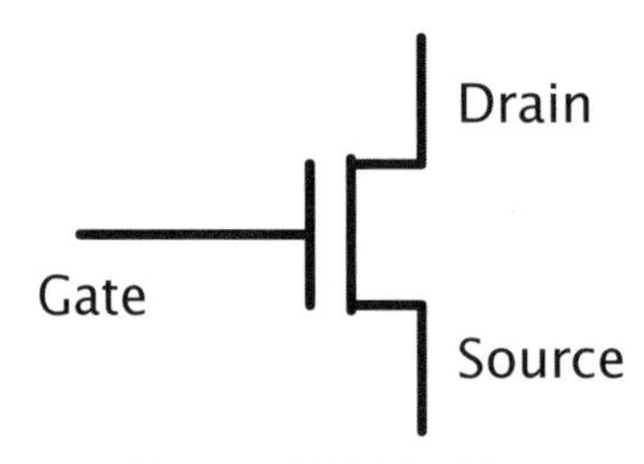

Figure 3-6 *The FET*

Virtually no current is needed at the gate to affect an FET; this makes it an ideal component for amplifying a signal that is weak, as the FET will not load the signal significantly. In fact, some of the better op-amps use FETs at their inputs for just this reason. One downside to an FET is that the parts tend to be easier to break than their transistor cousins. They are sensitive to static and over-voltage conditions, so be sure to pay attention to the maximum ratings when using these parts.

[4] You can buy a quad op-amp for less than 3 or 4 transistors these days, so why make it hard on yourself if you don't have to?

One very cool thing about an FET is the drain-to-source connection. It acts just like a resistor that you control by the voltage at the gate. This in effect makes it an electronically controlled variable resistor. Because of this, it is common to find FETs in circuits creating variable gain control. The drain-to-source connection acts like a resistor in either direction. That is, current can flow either way. However it is not uncommon for a FET to have a built-in reverse-biased diode across the drain-to-source pins. In this case, the output pins act like a resistor when this diode is reverse-biased and a diode when it is forward-biased.

When used in switch mode, a term you should pay attention to is *RDSon*. This is the Resistance Drain to Source when the device is turned all the way on. The lower this number, the less power you will lose across the device as heat. The voltage across the device will be the current times RDSon, and the power dissipated in heat will be this voltage times the current through the device.

An ohm equals volts divided by current if Ohm's Law still holds true (by this point in the book, a resounding YES should be on the tip of your tongue). The inverse of an ohm or 1/R equals current divided by voltage. This is known as a *mho*.[5] Mhos are to FETs as beta or HFE is to a transistor. This is the unit of gain, also known as transconductance that defines the output of the FET. Put X volts into the gate of the FET times that by the transconductance and you will get Y current drain to source.

Just as with transistors, this gain from input to output varies significantly. When using them in linear mode you need to either characterize the component you are using or develop some type of feedback control method that compensates for the variation to achieve the desired result.

In my experience, some engineers really like FETs and some like the good old BJT. I say, keep both tools in your tool chest, and use the right one for the job at hand.

[5] This unit is also know as a siemens, after that well-known brand name on many electronic gadgets you see around today. (OK, so it is really named after the guy who started that company that makes the stuff today.) Anyway, I like mho better; it just makes sense, since it is the inverse of an ohm after all. I have no idea as to the origin of the word mho, but if this book turns out successful enough to warrant a second edition, I hope that a reader smarter than me might clue me into the source of "mho."

Some Parts You Probably Haven't Heard Of

Here are a few parts in the semiconductor world that you may or may not have heard of.

Darlington transistor. This type of transistor consists of two transistors hooked together to increase the gain, as can be seen by the symbol used to represent it. Note that the base emitter diode drop is basically doubled in a Darlington transistor.

SCR. This is what you get when you create a PNPN junction, called a *silicon-controlled rectifier*. It is basically the combination of a diode and a transistor, it can switch large currents easily. But one caveat—you can turn it on but not turn it off. The current through the SCR must get below the holding current (very small) before it turns itself off. The SCR is part of the thyristor family.

TRIAC. This is a cousin to the SCR and also is in the thyristor family. Think of it as two SCRs back to back, making it an effective AC switch. It is often found in solid-state relays and such.

IGBT. Isolated gate bipolar transistor, this is best thought of as a combination between a Transistor and a FET. A FET is used to push a load of current through a big transistor.

There aren't really a lot of different variations in semiconductors; they all boil down to some basic configurations of the P and N materials. It is amazing to me that such a level of complexity is achieved from just a few parts, but semiconductors have truly revolutionized the world as we know it today. The devil is in the details, however. I can't stress too much the need to look at the datasheet of the part you are using. The more you know about its idiosyncrasies, the better your designs will be.

Power and Heat Management

One thing in common with all electrical devices (this side of superconductors) is the fact that, as they operate, heat is generated. This is because in every component (as we will learn later) there is some amount of equivalent resistance. Resistance times current flow equals a voltage drop, and a voltage drop times current equals power. Since Ohm's Law is unavoidable, this power must turn into heat. Heat is the premier cause of wear and tear in electronic components, so managing heat is a good thing to know something about. Let's start from the inside out.

Junction Temp

Inside a semiconductor, the place where all the magic happens is called the *junction*. This is the point where all the heat comes from as the part operates. The junction will have a maximum temperature that it can reach before something goes wrong.

Case Temp

The junction is always inside some type of case. Since you can't measure the junction temperature when you need to test out a design, you have to measure the case temperature. There will always be a temperature drop from the junction to case. The amount will typically be indicated in the spec sheet of the part. If the spec sheet says this case-to-junction thermal drop is 15°C, then you can expect the junction temp to be 15° warmer than what you measure. Here is a place where a good engineer will fudge the numbers in his favor. If the boss asks you to run this part as close to the edge as possible, you just tell him you need to be 30° under the junction temp as per the spec sheet. Since he most likely won't know where to look for this information, he will probably believe you and you will have a more robust design.

Heat Sinking

How hot the case gets depends on the heat sink attached to it. The case itself will be able to radiate a certain amount into the air around it. If this isn't sufficient, a heat sink can be added. One point you should recognize is that a heat sink (contrary to what you might think, given the name) is not a hole into which you can dump the heat from the part. A heat sink is more accurately described as a way to more efficiently transfer heat into the surrounding environment (the air in most cases).

Heat sinks capture that thermal rise and dissipate it into the surrounding air. Heat sinks are rated by a °C/W number. This number represents how much the temperature of the device on the sink will rise for every watt of heat generated. For example, if you put 20 watts of heat on a 3°C/W heat sink, the power device hooked up to that heat sink will rise 60°C above the ambient temperature.

Heat sinks can be thought of as heat conductors. Just as some metals are better electric conductors than others, some metals are better heat conductors. Usually one goes with the other. Aluminum is a better electrical conductor than steel, and it is also a better heat conductor. Copper, one of the best electrical conductors around, is also one of the best heat conductors. Thought of in these terms, the heat sink conducts heat away from the part. Like the fact that current always flows in one direction, heat always flows from hot to cold. There are a couple of ways for this to happen.

Radiation

Once the heat sink is warm, it will emit infrared radiation; as this energy is radiated away, the heat sink will cool. Have you ever wondered why so many heat sinks are black? This is because the color black[6] is an efficient radiator, just like this color tends to absorb more infrared radiation (as you may have noticed if you have ever worn a black shirt on a sunny day). It will radiate this heat away as well, as long as it is in a cooler environment and the sun isn't shining on it! While radiation is a way of getting heat moving, there are much better ways in most electronic devices today to get rid of heat.

Convection

The best way to get rid of heat is by moving some air across your heat sink. This is called *convection*. There are two ways to achieve convection: one by placing the heat sink so that air that is warmed by proximity to the heat sink rises. As this happens, cooler air takes its place to be warmed up and the whole process repeats. Most heat sinks have some type of spec as to free air operation that describes their function in this case.

---

[6] The color is not a major player when it comes to getting rid of heat, but it does help, so if you really need that last little bit of power handling, go black (but a little more metal will work just as well).

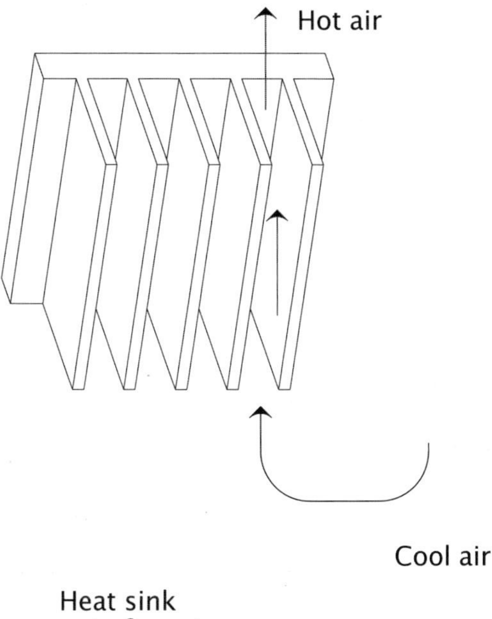

Figure 3-7 *Convection on a heat sink*

One quick side note—free air convection relies on the presence of gravity (hot air won't rise to be replaced by the cooler air without gravity), so if you happen to be working on a space shuttle experiment, don't count on free air convection for cooling!

A huge difference in cooling a heat sink can be achieved by moving more air across it. This is commonly accomplished by some type of fan. It is not unusual to see a heat sink handle 10 times as much power just by placing a fan next to it. This is the reason so many devices these days have acquired that proverbial hum that is so prevalent.

The more heat sink area you have in contact with the air, the better it can transfer heat. For this reason, you will see a lot of fins on these parts. More fins mean more surface area, which means more efficient heat transfer.

Hmmm, here's a thought—wouldn't it really be nice to recapture this heat and turn it back into power? I know there are thermoelectric devices that generate electricity when you heat them up, so this seems like a no-brainer…I guess I will get to that design later, but if any of you reading this get to the punch before me and make millions with this idea, all I ask is one percent!

Conduction

Another way of moving heat is by *conduction*. This is how the heat gets from the part into the heat sink, and it is how the heat travels across the sink as well. Conduction moves heat very, very well, but whatever it is conducting to must be cooler than where the heat is coming from in order for the heat to flow. Often a liquid is used to conduct heat away from stuff that gets hot, such as a nuclear reactor or your car engine. At the end of the day, though, that heat has to go somewhere. That is why you see a radiator in the front of your car dumping all that heat collected by the antifreeze into the atmosphere. The engine in my boat uses the entire lake as a heat sink, with no radiator needed, as it should be fairly obvious my piddling little boat isn't going to have enough power to raise the average temperature of millions of gallons of water by even a fraction of a degree.[7]

Can You Dump It Into a PCB?

This is a question that I have often heard. Can you use the PCB as a heat sink? The answer is yes. In fact, the PCB is simply copper plating and we know that copper is a good heat conductor, so it follows that it can be used as a heat sink. OK, here it comes... ***BUT...*** how do you know how well the PCB radiates the heat into the atmosphere? That is something you will most likely have to test to figure out. There are just so many variables in calculating this that it is faster to lay out the PCB, stick the part on and try it. Here are some items to note when using a PCB as a heat sink:

- A lot of little vias connecting the top layer to the bottom one will help increase the amount of surface area you have to dissipate the heat.

[7] You might even say, "Forget about the greenhouse effect, what about all this energy we are pouring into the atmosphere off our heat sinks?" If you consider on average every house in the world dumps 500W of heat from light bulbs alone into the atmosphere and figure about a billion houses, that comes out to a lot of energy! Is it enough to raise the temperature of the earth? I would have to dig a lot further back into chemistry classes than I would like to figure that out. But to simply spout generalities, sooner or later if we keep making more heat we will cook.

- The PCB in this area is going to get warm. That means expansion and contraction of the PCB. You may find this could cause mechanical damage over time, or even crack solder joints and PCB connections.
- I would recommend keeping the PCB heatsinks under 60°C. A cool rule of thumb I have learned is that if a metal surface is hot enough to burn you it is over 60°C.[8]

Heat Spreading

One of the major factors that control heat conduction when you have two materials next to each other is the surface area of the two materials that are touching. One other thing that affects conduction of a single material is the thickness of the material.

This gives rise to a technique known as *heat spreading*. A big, thick, very thermally conductive material is bolted up to the "hot part" to serve as a high-speed conduit to a bigger heat sink where all the fins for radiating the heat are located. The idea is to keep the junction temperature of the device lower by getting the heat away faster.

Does it work, you say? Truth is it can work, but there are many variables involved (such as the thermal conductivity between the heat spreader block and the rest of the heat sink, for example) that, as in the case of the PCB, will send you to the test lab to see if it is really working OK, or even helping. Remember, though, there will be a temperature gradient everywhere there is a junction between two parts; the fewer junctions, the better your heat sink will work.

[8] By no means am I endorsing touching a hot component as a way of checking its temperature! I hope that this disclaimer is enough to keep the sue-happy people out there off my case. I wouldn't want anyone to get burnt. I could go on about the legal ills that are crippling our world, but that is a whole other topic. Suffice it to say, if you happen to get burned on accident, you can be reasonably sure the metal you touched was over 60°C. Please don't touch it on purpose; there are much more accurate ways of measuring temperature than using your finger.

Thumb Rules

- Diodes are a "one way" valve for electrons.
- Diodes have a forward voltage drop you must overcome before it will conduct.
- Transistors are current driven.
- Transistors have a diode in the base that needs to be biased to work right.
- When using transistors as switches, check saturation current.
- FETs are voltage driven.
- FETs tend to be less robust; take care to design plenty of headroom between your circuit and the maximum ratings of the part.
- FETs are static sensitive.
- Study the datasheet of the part you are using meticulously.
- Heat is the biggest killer of electronic components.
- Most heat sinks dump heat into the air around them, most commonly by convection.
- If a part burns you when you touch it, it is over 60°C.
- You can use a PCB as a heat sink, but take care to test it.

The Magical Mysterious Op-Amp

Op-Amps—The Misunderstood Magical Tool!

Op-amps, in my opinion, are probably the most misunderstood yet potentially useful IC at the engineer's disposal. It makes sense that if you can understand this device you can put it to use, giving you a great advantage in designing successful products.

What Is an Op-amp Really?

Do you understand how an op-amp works? Would you believe that op-amps were designed to make it *easier* to create a circuit? You probably didn't think that the last time you were puzzling over a misbehaving breadboard in the lab.

In today's digital world, it seems to be common practice to breeze over the topic of op-amps, giving the student a dusting of commonly used formulas without really explaining the purpose or theory behind them. Then, the first time an engineer designs an op-amp circuit, the result is utter confusion when the circuit doesn't work as expected. This discussion is intended to give some insight into the guts of an operational amplifier, and to give the reader an intuitive understanding of op-amps.

One last point—make sure you read this section first! It is my opinion that one of the causes of op-fusion (op-amp confusion) as I like to call it, is that the theory is taught out of order. There is a very specific order to this, so please understand each section before moving on.

First, let's take the symbol of an op-amp.

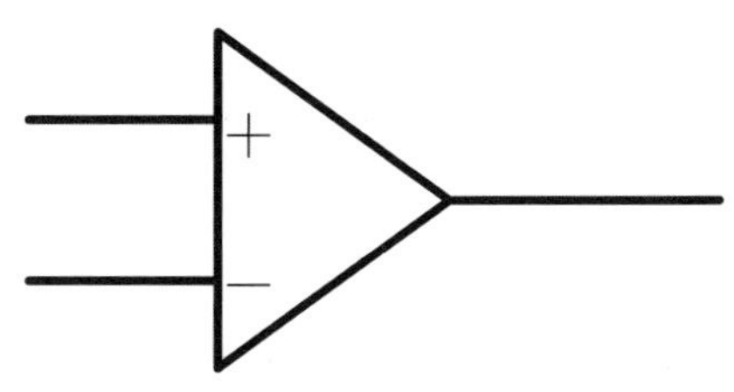

Figure 3-8 *Your basic op-amp*

There are two inputs, one positive and one negative, identified by the + and – signs.

There is one output.

The inputs are high impedance. I repeat. The inputs are high impedance. Let me say that one more time. THE INPUTS ARE HIGH IMPEDANCE! This means they have (virtually) no effect on the circuit to which they are attached. Write this down, as it is very important. We will talk about this in more detail later. This important fact is commonly forgotten and contributes to the confusion I mentioned earlier.

The output is low impedance. For most analysis it is best to consider it a voltage source.

Now let's represent the op-amp with two separate symbols:

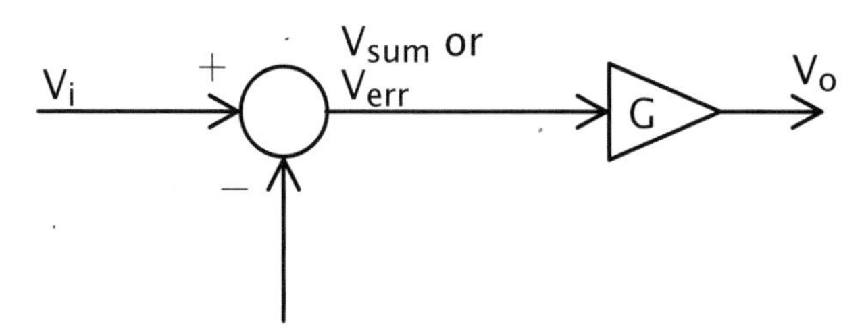

Figure 3-9 *What is really inside an op-amp*

You see here a summing block and an amplification block. You may remember similar symbols from your control theory class. Actually they are not just similar—they are exactly the same. Control theory works for op-amps. More on that later, too.

First, let's discuss the summing block. You will notice that there is a positive input and a negative input on the summing block, just as on the op-amp. Recognize that the negative input is as if the voltage at that point is multiplied by a –1. Thus, if you have 1V at the positive input and 2V at the negative input, the output of this block is –1. The output of this block is the sum of the two inputs, where one of the inputs is multiplied by –1. It can also be thought of as the difference of the two inputs and represented by this equation: $Vs = (V+) - (V-)$.

Now we come to the amplification block. The variable G inside this block represents the amount of amplification that the op-amp applies to the sum of the input voltages.

This is also known as the open-loop gain of the op-amp. In this case, we will use a value of 50,000. You say, how can that be? The amplification circuit I just built with an op-amp doesn't go that high! We will get to the amplification applications in a moment. Go find the open-loop gain in the manufacturer's datasheet. This level of gain or even higher is typical of most op-amps.

Now for some analysis. What will happen at the output if you put 2V on the positive input and 3V on the negative input. I recommend that you actually try this on a breadboard. I want you to see that an op-amp can and will operate with different voltages at the inputs. However, a little math and some common sense will also show us what will happen. For example:

$$Vout = 50{,}000 * (2 - 3), \text{ or } -50{,}000\text{V} \qquad \text{Eq. 3-1}$$

Unless you have a 50,000V op-amp hooked up to a 50,000V bipolar supply, you won't see –50,000V at the output. What will you see? Think about it a minute before you read on. The output will go to the minimum rail. In other words, it will try to go as negative as possible. This makes a lot of sense if you think about it like this. The output wants to go to –50,000V and obey the mathematics above. It can't get there, so it will go as close as possible. The rails of an op-amp are like the rails of a train track—a train will stay within its rails if at all possible. Similarly, if an op-amp is forced outside its rails, disaster occurs and the proverbial magic smoke will be let out of the chip. The rail is the maximum and minimum voltage the op-amp can output. As you can intuit this depends on the power supply and the output specifics of the op-amp.

OK, reverse the inputs. Now the following is true:

$$Vout = 50{,}000 * (3 - 2), \text{ or } +50{,}000\text{V} \qquad \text{Eq. 3-2}$$

What will happen now? The output will go to the maximum rail. How do you know where the output rails of the op-amp are? That depends on the power supply you are using and the specific op-amp. You will need to check the manufacturer's datasheet for that information. Let's assume we are using an LM324, with a +5V single-sided supply. In this case the output would get very close to 0V when trying to go negative and around 4V when trying to go positive.

At this time I would like to point something out. The inputs of the op-amp are NOT equal to each other. Many times I have seen engineers expect these inputs to be the same value. During the analysis stage, the designer comes up with currents going into the inputs of the device to make this happen (remember, high impedance inputs, virtually zero current flow). Then when he tries it out, he is confused by the fact that he can measure different voltages at the inputs.

In a special case we will discuss in Part 2, you can make the assumption that these inputs are equal. It is NOT the general case. This is a common misconception. You must not fall into this trap or you will not understand op-amps at all.

The examples above indicate a very neat application of op-amps: the comparator circuit. This is a great little circuit to convert from the analog world to the digital one. Using this circuit you can determine if one input signal is higher or lower than another. In fact, many microcontrollers use a comparator circuit in analog-to-digital conversion processes. Comparator circuits are in use all around us. How do you think the street light knows when it is dark enough to turn on? It uses a comparator circuit hooked up to a light sensor. How does a traffic light know when there is enough weight on the sensors to trigger a cycle to green? You can bet there is a comparator circuit in there.

Thumb Rules

- The inputs are high impedance; they have negligible effects on the circuit they are hooked to.
- The inputs can have different voltages applied to them; they do NOT have to be equal.
- The open-loop gain of an op-amp is VERY high.
- Due to the high open-loop gain and the output limitations of the op-amp, if one input is higher than the other the output will "rail" to its maximum or minimum value (this application is often called a *comparator circuit*).

Negative Feedback

If you didn't just finish reading it, go back and read the thumb rules from the last section. They are very important to develop the correct understanding of what an op-amp does. Why are these points important? Let's go over a little history. Up until the invention of op-amps, engineers were limited to the use of transistors in amplification circuits. The problem with transistors is that, being "current-driven" devices, they always affect the signal of the circuit that the designer wants to amplify by loading the circuit. Due to manufacturing tolerances of transistors, the gain of the circuits would vary significantly. All in all, designing an amplifier circuit was a tedious process that required much trial and error. What engineers wanted was a simple device that they could attach to a signal that could multiply the value by any desired amount. The device should be easy to use and require very few external components. To paraphrase, *operation* of this *amplifier* should be a "piece of cake." At least that is the way I remember it. The other way the name *operational amplifier* or op-amp came into being was to describe the fact that these amplifiers were used to create circuits in analog computers, performing such *operations* as multiplication, among others.

To begin with, let's take a look at the special case I mentioned in the previous discussion. First, return to the previous block diagram and add a feedback loop.

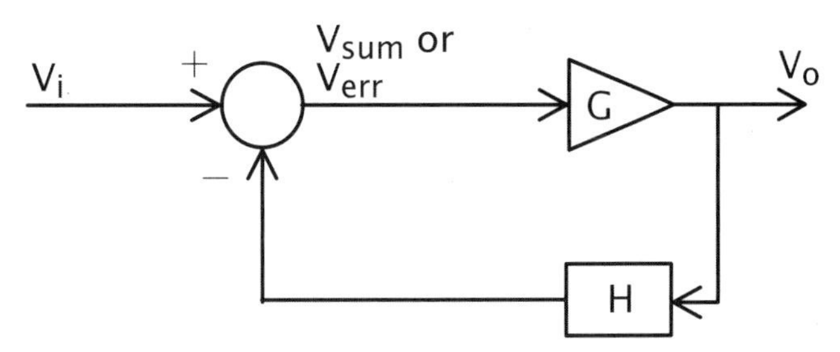

Figure 3-10 *Adding negative feedback to the op-amp*

You will see that I have represented the forward or open-loop gain with the value G, and the feedback gain with the value *H*. (This diagram should look very familiar to those of you who have had training in control theory.) First, you see that the output is tied to the negative input. This is called *negative feedback*. What good is negative feedback? Let's try an experiment. Hold your hand an inch over your desk, and keep it there. You are experiencing negative feedback right now. You are observing via sight and feel the distance from your hand to the desk. If your hand moves, you respond with a movement in the opposite direction. That is negative feedback. You invert the signal you receive via your senses and send it back to your arm. The same thing occurs when negative feedback is applied to an op-amp. The output signal is sent back to the negative input. A signal change in one direction at the output causes a V_{sum} to change in the opposite direction.

You should get an intuitive grasp of this negative feedback configuration. Look at the previous diagram and assume a value of 50,000 for G and a value of 1 for *H*. Now start by applying a 1 to the positive input. Assume the negative input is at 0 to begin with. That puts a value of 1 at the input of the gain block G and the output will start heading for the positive rail. But what happens as the output approaches 1? The negative input also approaches 1. The output of the summing block is getting smaller and smaller. If the negative input goes higher than 1, the input to the gain block G will go negative as well, forcing the output to go in the negative direction. Of course, that will cause a positive error to appear at the input of the gain block G, starting the whole process over again. Where will this all stop? It will stop when the negative input is equal to the positive input. In this case since *H* is 1, the output will be 1 also.

You have learned this in control theory. Look at the basic control equation in reference to the previous diagram:

$$V_o = V_i * \frac{G}{1 + G * H} \qquad \text{Eq. 3-3}$$

What happens when G[9] is very large? The 1 in the denominator becomes insignificant and the equation becomes V_o = approximately V_i * (1 / *H*). *H* in this case is 1 so it follows that V_o = approximately V_i * (1 / 1) or,

$$V_o = V_i \qquad \text{Eq. 3-4}$$

This is the special case where you can assume that the inputs of the op-amp are equal. Apply it ONLY when there is negative feedback. When feedback gain is one, this also demonstrates another neat op-amp circuit, the voltage follower. Whatever voltage is put on the positive input will appear at the output.

Take a look at the following figure. This is an op-amp in the negative feedback configuration. When you look at this, you should see a summer and an amplifier just as in the previous drawing. In this configuration, you can make the assumption that the positive and negative inputs are equal.

Negative feedback is the case that is drilled into you in school, and is the one that often causes confusion. It is a special case, a very widely used special case. Nonetheless, if you do not have negative feedback and the inputs and output are within operational limits, you must NOT assume the inputs of the op-amp are equal.

[9] Remember an op-amp has a very large "G"!

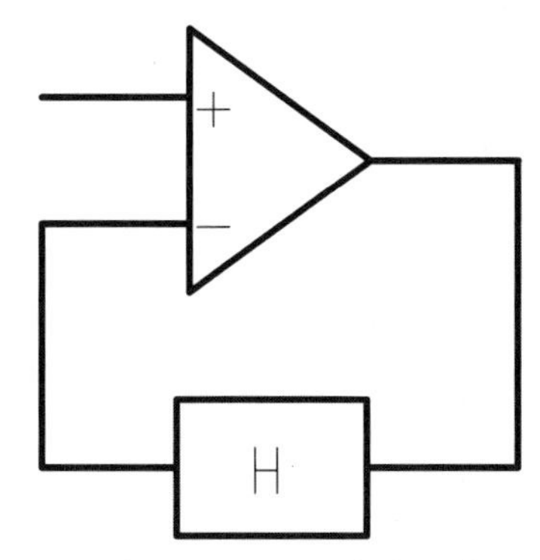

Figure 3-11 *Original op-amp symbol with negative feedback*

Why is this negative feedback configuration used so much? Remember the reason op-amps were invented? Amplifiers were tough to make. There had to be an easier way. Take a look at the control equation again:

$$V_o = V_i * \frac{G}{1 + G * H} \qquad \text{Eq. 3-5}$$

I have already shown that for large values of G, the equation approximates:

$$V_o = V_i * \frac{1}{H} \qquad \text{Eq. 3-6}$$

You will see that the amplification of V_i depends on the value of H. For example, if we can make H equal 1/10, then $V_o = V_i * (1 / (1 / 10))$ or,

$$V_o = V_i * 10 \qquad \text{Eq. 3-7}$$

How do we go about doing that? Do you remember the voltage divider circuit? That would be very useful here, as we would like H to be the equivalent of dividing by 10. Lets insert the voltage divider circuit in place of H. (Note, V_i will be connected to V+.)

Notice that the input to the voltage divider comes from the output of the op-amp, V_o. The output of the voltage divider goes to the negative input of the op-amp $V–$. Now, will the op-amp input V– affect the voltage divider circuit? NO! It has high impedance. It will not affect the divider. (If you didn't get that, go back and read "what's in an op-amp really" till you do!) Since the input to the divider is hooked to a voltage source, and the output is not affect by the circuit, we can calculate the gain from V_o to $V–$ very easily with the voltage divider rule.

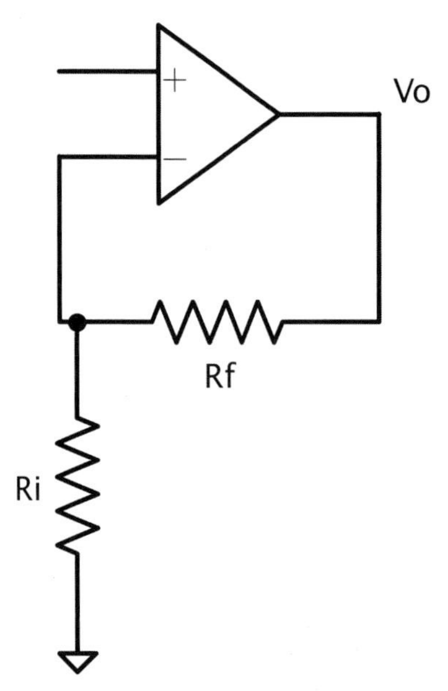

Figure 3-12 *Negative feedback is a voltage divider*

$$\frac{V-}{V_o} = \frac{R_i}{R_i + R_f} = H \qquad \text{Eq. 3-8}$$

Thus it follows that:

$$\frac{1}{H} = \frac{R_i + R_f}{R_i}$$, or with a little algebra,

$$\frac{1}{H} = \frac{R_i}{R_i} + \frac{R_f}{R_i} = \frac{R_f}{R_i} + 1 \text{ or } \frac{1}{H} = \frac{R_f}{R_i} + 1 \qquad \text{Eq. 3-9}$$

There you have it—the gain of this op-amp circuit. Let's look at it another way. Go back to the previous equation:

$$\frac{V-}{V_o} = \frac{R_i}{R_i + R_f} \qquad \text{Eq. 3-10}$$

We learned that in this special case of negative feedback we can assume that V+ = V–. This is because the negative feedback loop is pushing the output around, trying to reach this state. So let's assume that V_i = V+ which is where the input to our amplifier will be hooked up. Now we can replace V– with V_i, and the equation looks like this:

$$\frac{V_i}{V_o} = \frac{R_i}{R_i + R_f} \qquad \text{Eq. 3-11}$$

What we really want to know is what does the circuit do to V_i to get V_o? Let's do a little math to come up with this equation:

$$V_o = V_i * \frac{R_i + R_f}{R_i} = V_i * \frac{R_f}{R_i + 1} \text{ or } \frac{V_o}{V_i} = \frac{R_f}{R_i} + 1 \qquad \text{Eq. 3-12}$$

Please note that this is equal to $1/H$. You see, the gain of this circuit is controlled by 2 simple resistors. Believe me, that is a whole lot easier to understand and calculate than a transistor amplification circuit. As you can see, the operation of this amplifier is pretty easy to understand.

Thumb Rules

- The negative feedback configuration is the only time you can assume that $V- = V+$.
- The high impedance inputs and the low impedance output make it easy to calculate the effects simple resistor networks can have in a feedback loop.
- The high open-loop gain of the op-amp is what makes the output gain of this special case equal to approximately $1/H$.
- Op-amps were meant to make amplification easy, so don't make it hard!

Positive Feedback

What is positive feedback? Let's take a look at a real-world example. You are working one day, and your boss stops by and says, "Hey, you should know that you've handled your project very well, and that new op-amp circuit you built is awesome!" After you bask in his praise for a while, you find yourself working even harder than before. That is positive feedback. The output is sent back to the positive input, which in turn causes the output to move further in the same direction. Let's look at the diagram of an op-amp again.

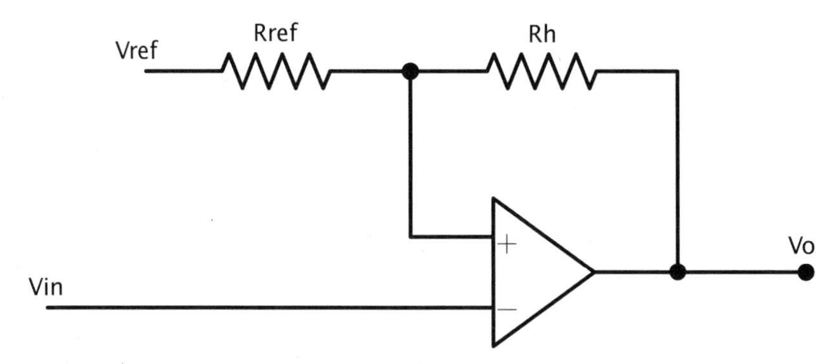

Figure 3-13 *Positive feedback on an op-amp*

Now we will do a little intuitive analysis. Don't forget the thumb rules we learned in the last two sections. Review them now if you need to.

First, apply 0V to V_{in}. In this case the input is connected to V–. You also see that the output is connected via a resistor to a reference voltage, V_{ref}. What is the voltage at V+? Does the voltage at V+ equal the voltage at V–? NO! (Don't believe me? Check the thumb rules!)

What *is* the voltage at V+? That depends on two things: the voltage at V_{ref} and the output voltage of the amplifier, Vo. Does the V+ input load the circuit at all? No, it does not. To begin the analysis, let V_{ref} = 2.5V, and assume the output is equal to 0V. Now what is the voltage at V+? What do you know, since V_o is equal to 0, we have a basic voltage divider again. Assume R_{ref} = 10K and R_h = 100K:

$$V+ = V_{ref} * \frac{R_h}{R_h + R_{ref}} = 2.5 * \frac{100K}{110K} = 2.275V \qquad \text{Eq. 3-13}$$

So now there is 2.275V at V+ and 0V at V–. What will the op-amp do? Let's refer to the block diagram of the op-amp we learned earlier.

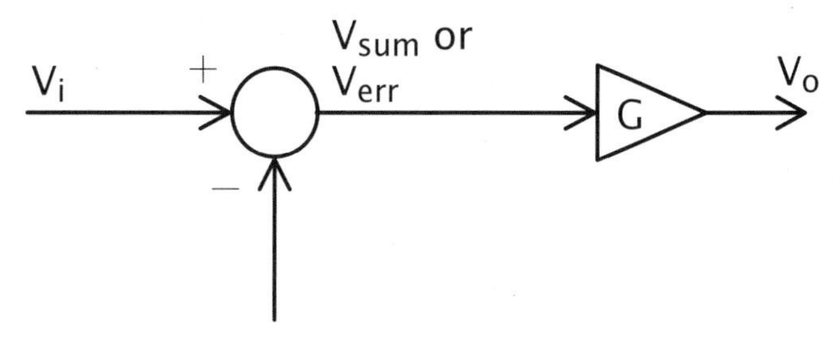

Figure 3-14 *Start with what is really inside!*

What do we have? V_{sum} is equal to V+ – V– or, in this case, V_{sum} = 2.275V. V_o is equal to V_{sum} * G. The output will obviously go to the positive rail (if this is not obvious to you, you need to review "what is an op-amp really" again). Now we have V_o at the positive rail. Let's assume that it is 4V for this particular op-amp. (Remember, the output rails depend on the op-amp used, and you should always refer to the datasheets for that information. 4V used in this case is typical for an LM324 with a 0 to 5V supply.)

The output is at 4V and V– is at 0V, but what about V+? It has changed. We must go back and analyze it again. (Do you feel like you are going in circles? You should. That is what feedback is all about; outputs affect inputs which affect the outputs, and so on, and so on.) The analysis this time has changed slightly. It is no longer possible to use just the voltage divider rule to calculate V+. We must also use superposition.

In superposition, you set one voltage source to 0 and analyze the results, and then you set the other source to 0 and analyze the results. Then you add the two results together to get the complete equation. Let's do that now. We already know the result due to V_{ref} from above. Here is the positive feedback diagram again for reference.

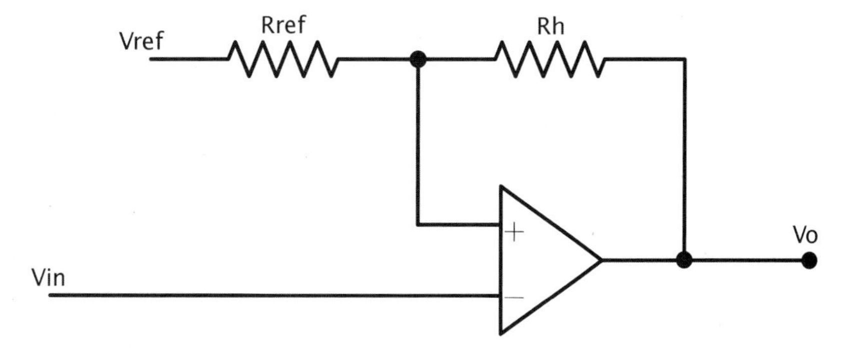

Figure 3-15 *Positive feedback on an op-amp*

Here is the result due to V_{ref} using the voltage divider rule:

$$V+ \text{ due to } V_{ref} = \frac{V_{ref} * R_h}{R_h + R_{ref}} \qquad \text{Eq. 3-14}$$

Here is the result due to V_o using the voltage divider rule.

$$V+ \text{ due to } V_o = \frac{V_o * R_{ref}}{R_{ref} + R_h} \qquad \text{Eq. 3-15}$$

The result due to both is thus:

$V+ = (V+ \text{ due to } V_{ref}) + (V+ \text{ due to } V_o)$ or,

$$V+ = \frac{V_{ref} * R_h}{R_h + R_{ref}} + \frac{V_o * R_{ref}}{R_h + R_{ref}} \qquad \text{Eq. 3-16}$$

Now insert all the current values and we have:

$$V+ = \frac{2.5 * 100K}{110K} + \frac{4 * 10K}{110K} = 2.64V \qquad \text{Eq. 3-17}$$

Is this circuit stable now? Yes, it is. We have 0V at V–, and 2.64V at V+. This results in a positive error which, when amplified by the open-loop gain of the op-amp, causes the output to go to the positive rail. This is 4V, which is the state that we just analyzed.

Let's change something. Let's start slowly ramping up the voltage at V–. At what point will the op-amp output change? Right after the voltage at V– exceeds the voltage at V+. This results in a negative error, causing the output to swing to the negative rail. And what happens to V+? It changes back to 2.275V as we calculated above. So how do we get the output to go positive again? We adjust the input to less than 2.275V. The positive feedback reinforces the change in the output, making it necessary to move the input farther in the opposite direction to affect another change in the output.

The effect that I have just described is called *hysteresis*. It is an effect very commonly created using a positive feedback loop with an op-amp. What is hysteresis good for, you ask. Well, heating your house for one thing. It is hysteresis that keeps your furnace from clicking on and off every few seconds. Your oven and refrigerator use this principle as well. In fact, the disk drive on the computer I used to write this uses hysteresis to store information.

One important item to note. The size of the hysteresis window depends on the ratio of the two resistors R_{ref} and R_h. In most typical applications R_h is much larger that R_{ref}. If the signal at V_i is smaller than the window, it is possible to create a circuit that latches high or low and never changes. This is usually not desired and can be avoided be performing the analysis above and comparing the calculated limits to the input signal range.

Now that we have covered the three basic configurations of an op-amp, let's put together a simple circuit that uses them. Here we have a voltage follower, hooked to a comparator using hysteresis, with an LED as an indicator.

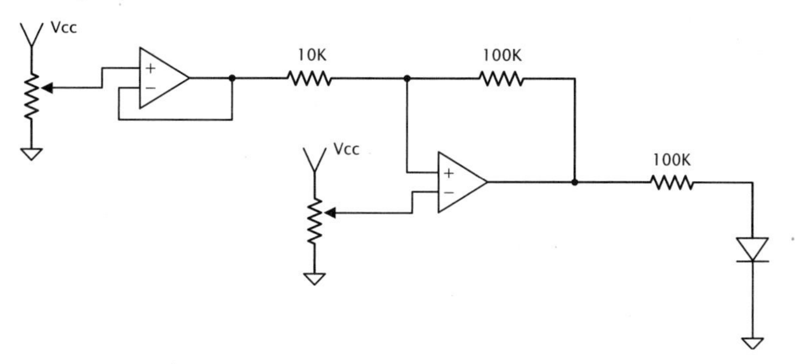

Figure 3-16 *Simple op-amp circuit for your bench to understand both positive and negative feedback*

You should build this in your lab to gain an intuitive understanding of what has been discussed. Experiment with feedback changes in all parts of the circuit. Note that you can change the input potentiometers from 5K to 100K without affecting the voltage at which the comparator switches.

All About Op-Amps

There you have it, the basics of op-amp circuits. With this information, you can analyze most op-amp circuits you come across and build some really neat circuits yourself. What about filters, you say! Well, a filter is nothing more than an amplifier that changes gain depending on the frequency. Simply replace the resistors with an impedance and thus add a frequency component to the circuit. What about oscillators, you say? These are feedback circuits where timing of the signals is important. They still follow the rules above. I believe that grasping the basics of any discipline is the most important thing you can do. If you understand the basics, you can always build on that foundation to obtain higher knowledge, but if you do not "get the basics" you will flounder in your chosen field.

Thumb Rules

- Op-amp inputs are high impedance (that means no current flows into the inputs); this can't be said too much so forgive me for repeating it.
- Op-amp outputs are low impedance.
- V+ = V− only if negative feedback is present, they don't have to equal if feedback is positive.
- Positive feedback creates hysteresis when properly set up.
- Positive feedback can make an output latch to a state and stay there.
- Positive feedback with a delay can cause an oscillation.
- Op-amps were designed to make it easy, so don't make it hard!

It's Supposed to Be Logical

Binary Numbers

Binary numbers are so basic to electrical engineering that I nearly omitted this section on the premise that you would already know it. However, my own words, "drill the basics," kept haunting me. So if you already know this stuff, you are authorized to skip this section, but if those same words start to haunt you like I hope they will, you should at least skim through it.

Binary numbers are simply a way to count with only two values, 1 and 0—convenient numbers for reasons we will discuss later. Binary is also known as base 2. There are other bases, such as base 8 (octal) and base 16 (hexadecimal) that are often used in this field, but it is primarily for the reason that they represent binary numbers easily. The common base that everyone is used to is decimal, also known as base 10. Think of it this way: the base of the counting system is the point at which you move a digit into the left column and start over at 0. For example, in base 10 you count 123...789 and then you chalk one up in the left column and start over at zero for the number 10. In base 8 you only get to 7 before you have to start over ...5,6,7,10,11 and so on. Base 16 starts over at 15 in the same way, but to adhere to the rule of one digit in the column before we roll over into the next digit, we use letters to represent 10 through 15. A table is an easy way to see this relationship.

Table 3-1 *Decimal and hexadecimal numbers*

Decimal, Base 10	*Hexadecimal, Base 16*
0	*0*
1	*1*
2	*2*
3	*3*
4	*4*
5	*5*
6	*6*
7	*7*
8	*8*
9	*9*
10	*A*
11	*B*
12	*C*

(continued)

13	*D*
14	*E*
15	*F*
16	*10*
17	*11*
And so on...	

Please notice again how the numbers start over at the corresponding base. You might also notice that I started at zero in the counting process. It should be stressed that zero is an important part of any counting system, a fact that I think tends to get overlooked. If you think about it, if 0 is included, the point at which base 10 rolls over is the 10th digit and the point at which base 8 rolls over is the 8th digit. The same relationship exists for any base number you use.

So let's get back to binary or base 2. The first time I saw binary numbers I thought, "Wow, what a tantalizing numeric system. Just as soon as you make one move to get where you are going, it is time to start over again." The numbers go like this: 0, 1, 10, 11, 100… Again I think a table is in order.

Table 3-2 *Decimal, binary, octal and hexadecimal number comparison*

Decimal Base 10	***Binary Base 2***	***Octal Base 8***	***Hexadecimal Base 16***
0	*0*	*0*	*0*
1	*1*	*1*	*1*
2	*10*	*2*	*2*
3	*11*	*3*	*3*
4	*100*	*4*	*4*
5	*101*	*5*	*5*
6	*110*	*6*	*6*
7	*111*	*7*	*7*
8	*1000*	*10*	*8*
9	*1001*	*11*	*9*
10	*1010*	*12*	*A*
11	*1011*	*13*	*B*
12	*1100*	*14*	*C*
13	*1101*	*15*	*D*
14	*1110*	*16*	*E*
15	*1111*	*17*	*F*
16	*10000*	*20*	*10*
17	*10001*	*21*	*11*
18	*10010*	*22*	*12*
And so on...			

Notice how base 8 and base 16 roll over right at the same point that the binary numbers get an extra digit. That is why they are convenient to use in representing binary numbers. You may also have noticed that decimal numbers don't line up as nicely.

Another pattern you should see in this table is that you hit 20 in base 8 at the same point you see 10 in base 16. This makes sense because one base is exactly double the other. Can you extrapolate what base 4 might do?

This leads to another trick with binary numbers. Each significant digit doubles the value of the previous one (just like every digit you add in decimal is worth 10 times the previous one). Let's look at yet another table.

Table 3-3 *Doubling digits*

Decimal	*128*	*64*	*32*	*16*	*8*	*4*	*2*	*1*
Binary	*10000000*	*1000000*	*100000*	*10000*	*1000*	*100*	*10*	*1*

You can add up the values of each digit where you have a 1 in binary to get the decimal equivalent. For example, take the binary number 101. There is a 1 in the 1's column and in the 4's column. Add 1 plus 4 and you get 5, which is 101 in binary. You might also notice the numbers you can represent double for every digit you add to the number. For example, 4 digits let you count to 15, and 8 digits will get you to 255. (This causes some of the more extroverted engineers to attempt to become the life of the party by showing their friends they can count to 1023 with the fingers on their hands. The attempt usually fails.)

All the math tricks you learned with decimal numbers apply to binary as well, as long as you consider the base you are working in.

For example, when you multiply by 10 in decimal you just put a zero on the end, right? The same idea applies to binary, but the base is 2, so to multiply by 2, you just stick a zero on the end, shifting everything else to the left. When dividing by 10 in decimal you just lop off the last digit, and keep whatever was there as a remainder. Dividing by 2 in binary works the same way, shifting everything to the right, but the remainder is always 0 or 1—a fact that is convenient for math routines, as we will learn later.

For whatever reason, most electronic components like to manage binary numbers in groups of 4 digits. This makes hexadecimal or hex numbers a type of shorthand for referring to binary numbers. It is a good short hand to know.

In the electronics world, each binary digit is commonly referred to as a bit. A group of eight bits is called a *byte* and four bits is called a *nibble*. So if you "byte" off more than you can chew, maybe you should try a "nibble" next time.

So back to the point, since a hex number nicely represents a nibble, and there are two nibbles in a byte, you will often see two hex numbers used to describe a byte of binary information. For example, 0101 1111 can be described as 5 F or 1110 0001 as E 1. In fact you can easily determine this by looking up the hex equivalent to any nibble using Table 11-2.

To sum things up, binary numbers are a way to count using only two symbols; they are commonly referred to using hex numbers as a type of shorthand notation. When logic circuits came along, the fact that they represented information with only two symbols, on or off, high or low, made them dovetail nicely with binary numbers and binary math.

Logic

One of the most incredible growth industries over the last 50 years has come from the application of electronics to manipulate data based on the principles of Boolean logic. Originally developed by George Boole in the mid-1800s Boolean logic is based on a very simple concept, yet allows creation of some very complex stuff.

Let the value "1" mean true, and let the value "0" mean false. In an actual circuit, 1 may typically be any signal between 3–5V, and 0 any signal between 0–2.9V, but what is important in the world of logic is that there are only two states, 1 or 0. The world is black or white. That said, it is no wonder that engineers have so quickly grasped the digital domain. I haven't met an engineer who doesn't like his world to follow nice predictable rules. "Keep it simple" is a common mantra and resolving the world into two states sure does simplify things. It is important to note that at some point in the circuit a decision needs to be made whether the current value represents a 1 or a 0.

During our study of logic we will refer to a description of logic inputs and outputs known as truth tables. The inputs are generally shown on the left and the outputs are on the right. Some basic components that manipulate logic are called gates. Let's start with these basics.

The NOT Gate

This is as simple as it gets. The NOT gate inverts whatever signal you put into it. Put in a 1, get a 0 out, and vice versa. Lets take a transistor and make a NOT gate.

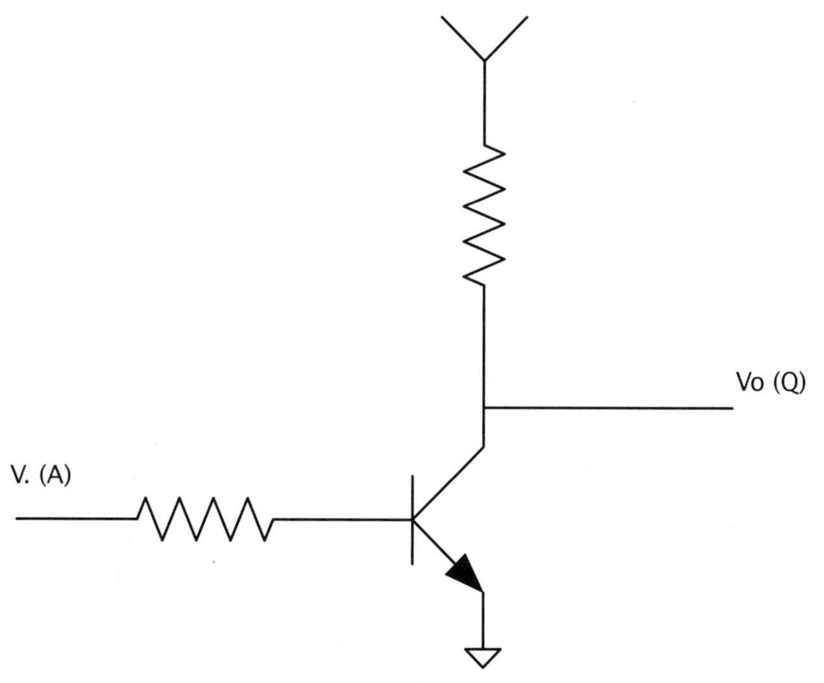

Figure 3-17 *Transistor NOT gate*

If you put 0V into this, you will get 5V out. If you put 5V into this, you will get nearly[10] 0V out. You have effectively inverted the logic symbol. The NOT gate, also called the *inverter*, is commonly represented by this symbol.

[10] Please note, I said *nearly* 0 volts. The output of this circuit does not get quite all the way to 0, but that doesn't matter as long as the value is below the maximum level for a 0.

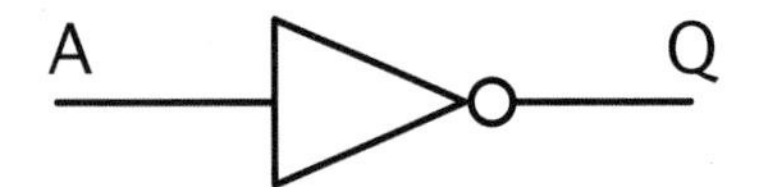

Figure 3-18 *Inverter or NOT symbol*

Here it the truth table:

Input A	Output Q
1	0
0	1

The AND Gate

The AND function is describe by the rule that all inputs need to be true or 1 in order for the output to be true. If this is true and that is true, then this AND that must be true. However, if either is false, then the output must be false. It is defined by the following truth table:

Input A	Input B	Output Q
0	0	0
0	1	0
1	0	0
1	1	1

We can build this circuit with only a couple of diodes.

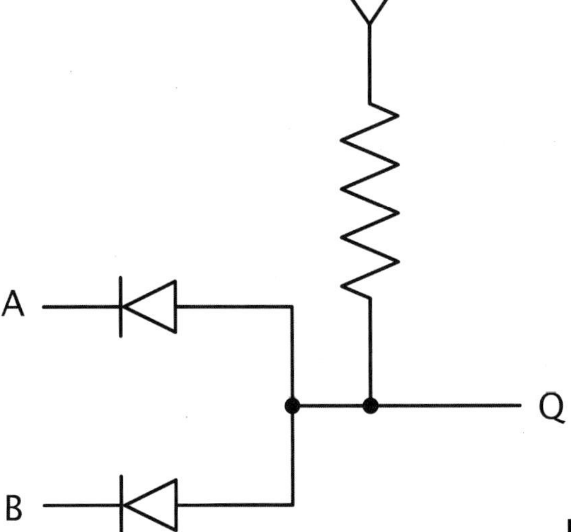

Figure 3-19 *Diode AND gate*

One way to think of it is if either input is false, then the output will be false. This function is commonly referred to by this symbol.

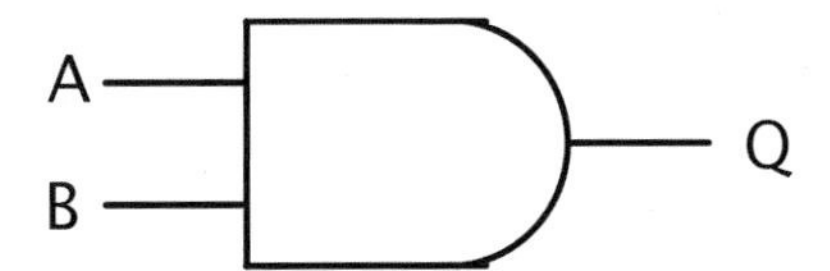

Figure 3-20 *Diode AND gate*

The OR Gate

Did you notice that three of the input conditions on the AND gate resulted in a false or 0 at the output? The OR gate is sort of the opposite, but not exactly. Three of the input conditions result in a true at the output, while only one condition creates a 0. If *this* is true OR *that* is true, then it only takes one true input to create a true output. Here is the truth table:

Input A	*Input B*	*Output Q*
0	0	0
0	1	1
1	0	1
1	1	1

We can make this circuit with diodes too; we just flip them around.

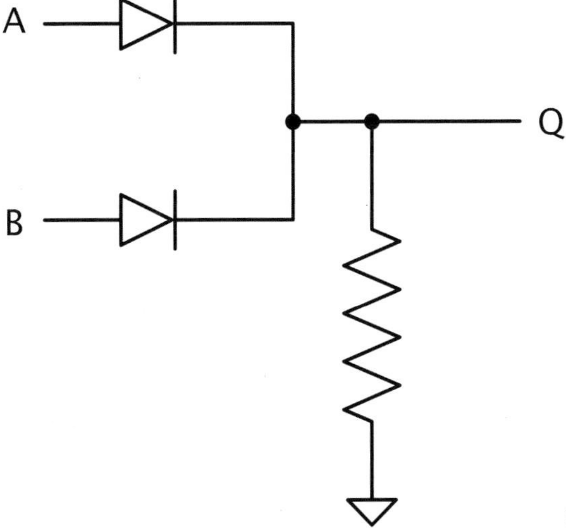

Figure 3-21 *Diode OR gate*

The more common OR symbol looks like this:

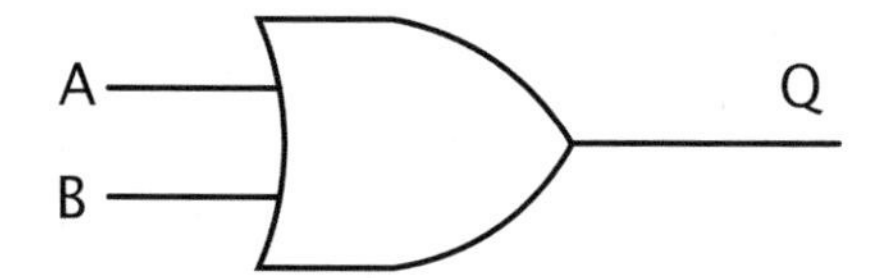

Figure 3-22 *OR symbol*

That's it—those are the basic gates. There are only three of them. "Now wait a minute," you may be saying, there were a lot more when I had logic circuits in class, weren't there? There are more gates, but they are all built from these three basic gates. If you understand these, you can derive the rest. With that in mind, see if you can make these other logic gates using only the three components above.

The NAND gate

NAND means NOT AND, and it is what it says. Invert the output of an AND gate with the NOT gate and you have a NAND gate. Here is the truth table.

Input 'A'	Input 'B'	Output 'Q'
0	0	1
0	1	1
1	0	1
1	1	0

Let's build one with the basic symbols we have already learned.

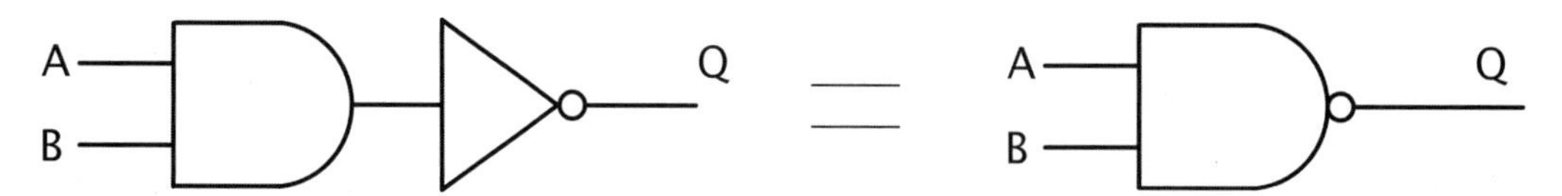

Figure 3-23 *How to build a NAND gate*

This gate is so commonly used that it has its own symbol. Note the little bubble on the output, which is used to indicate an inverted signal.

Can you make this with basic semiconductors as well? The answer is yes. In fact you only need two transistors.

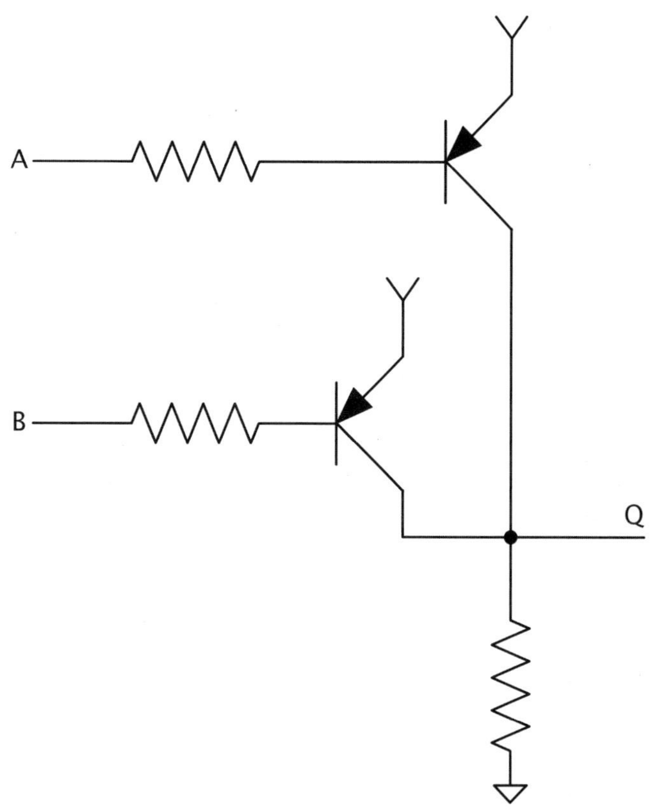

Figure 3-24 *Simple transistor NAND gate*

The NOR Gate

Yep, you guessed it, this is the NOT OR gate. It is made by inverting the output of the OR gate, just like the NAND gate. Here is the truth table:

Input A	*Input B*	*Output Q*
0	0	1
0	1	0
1	0	0
1	1	0

The NOR gate is an inverted OR gate with a symbol like this:

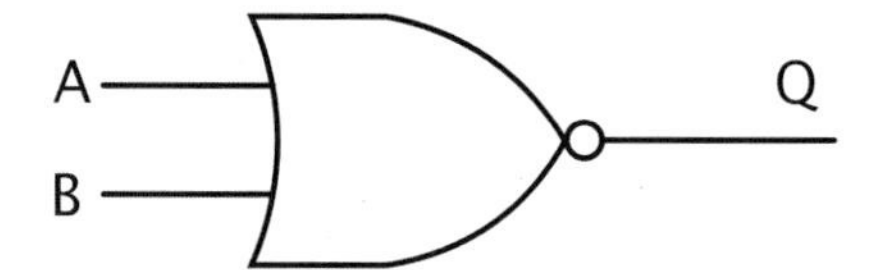

Figure 3-25 *NOR gate symbol*

Better yet, you can make this gate with only two transistors as well.

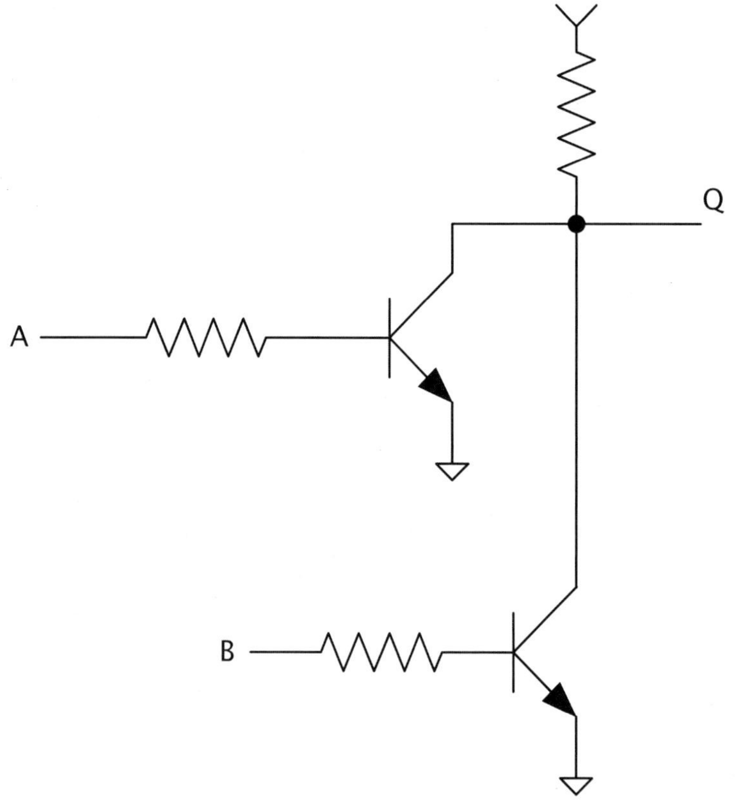

Figure 3-26 *Transistor NOR gate*

The XOR Gate

XOR means *exclusive or*. In words, think of it like this—it's true if *this* or *that* is true, but not if both are true. Following is the truth table:

Input A	***Input B***	***Output Q***
0	*0*	*0*
0	*1*	*1*
1	*0*	*1*
1	*1*	*0*

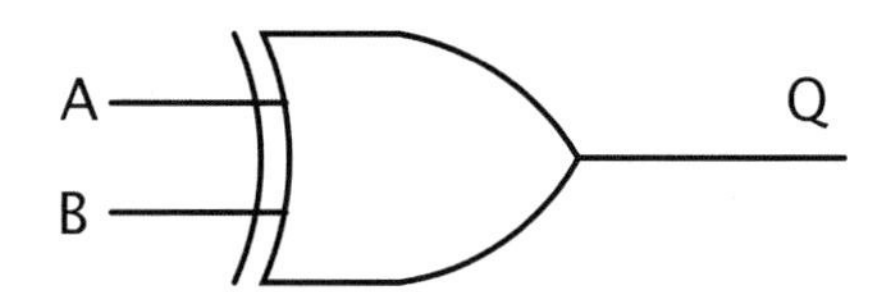

Figure 3-27 *XOR (exclusive OR) gate*

Let's see if we can make this with basic semiconductor components as we did with the other logic circuits.

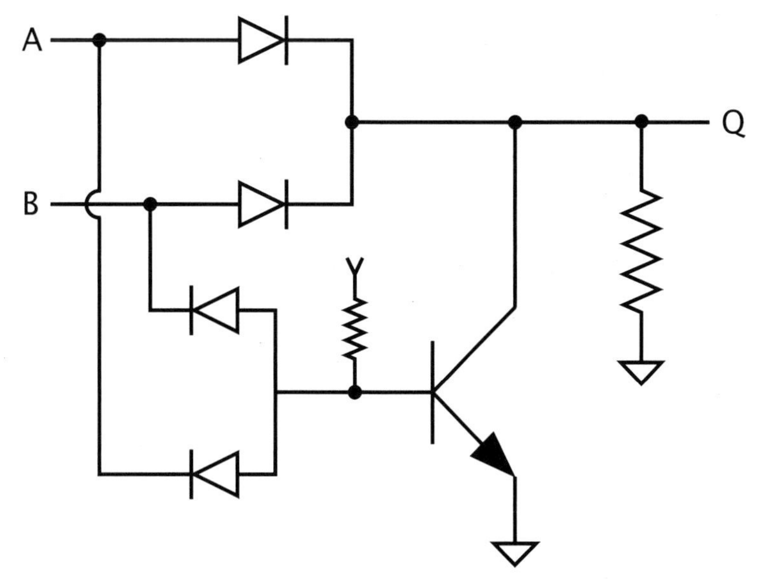

Figure 3-28 *Diode and transistor-based XOR gate*

The XNOR gate looks like this:

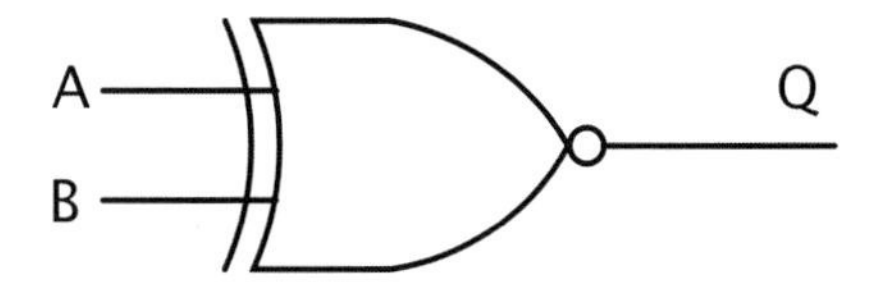

Figure 3-29 *The XNOR gate*

If I have done a good job with my explanations, the function of this gate should be obvious. It is an XOR with an inverted output. Here is its truth table:

Input A	*Input B*	*Output Q*
0	0	1
0	1	0
1	0	0
1	1	1

Adders

As you already know, it is possible to count with these ubiquitous 1s and 0s. The logical extension of counting is math!

Joining several of these gates together, we can create a binary adder; string a bunch of these adders together to add any number of binary digits and, since any number can be represented by a string of those pesky 1s and 0s, we now have the basis of computation. Are you beginning to see how that calculator[11] on your desk works?

Memory Cells

It is possible to use these devices to create what is called a memory cell. Here is a diagram of one.

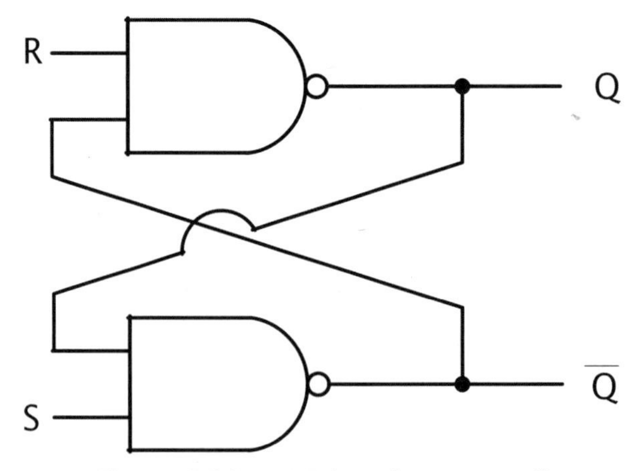

Figure 3-30 *NAND based memory cell*

[11] Technically, most calculators use a cordic algorithim. It is a slick way to handle things like sine, cosine and other stuff and still keep the electronics simple. At the end of the day though, there are still logic elements deep down inside that desktop appliance doing all the work.

The basic premise is that the cell will retain the state you set it to. Some memory will lose the data that was stored if power is lost; this is known as *volatile memory*. this is like the RAM in your computer. Another category of memory is known as *nonvolatile*. In this type the data is retained even when power is removed. An example of this is flash memory, commonly found in the now-ubiquitous thumb drive.

Now that you have the ability to make a decision, compute mathematical functions and remember the results to make more decisions later, you have the basics of a Turing machine. Alan Turing was a cryptographer who laid much of the foundation for computational theory. He described the Turing machine, a system that has an infinite amount of memory, the ability to go back and forth along that memory, and follow the instructions at any location. Aside from infinite memory, today's computers are as close as anything comes to a Turing machine.

From the simple gates that started it all to supercomputers, ever more complex systems are based on these simple logic components. It is no wonder that every new megacool processor has a gagillion transistors in it. There is a sort of "in-between" device that is worth mentioning, though, as it will help you grasp the complexities such a simple device can create. It is known as a *state machine*.

State Machines

State machines lie in the realm between discrete logic and a microcontroller. They usually have a clock of some type, memory, and most of the basic parts a micro has; however, they don't need all of these parts to operate.

As the name implies, the output of a state machine is a function of the "state" of the inputs at any given moment in time. Often a clock signal of some type is used to determine the moment that these inputs should be evaluated. Memory cells, also called *flip-flops*, are used to store information. A flip-flop reflects the state of the input at the time a clock signal was present. Thus, conditions used for evaluation can be stored in memory.

The inputs of a logic element can be detected at three different points in time on the clock signal, falling edge, rising edge, or level detect. Which one is used depends on the part itself; you will need to check that source of all knowledge, the datasheet.

These terms are self-explanatory: data is assessed when the clock signal rises, falls, or remains level. This makes timing of the signals important. This importance of timing will come up again as we explore microcontrollers (which are really just hopped-up state machines with a defined group of instructions, but more on that later).

Due to the falling cost of microcontrollers, I believe purely implemented state machines are going out of fashion these days. When they do appear, they are usually in a programmable logic device (PLD). Gone are the days of soldering a slew of D flip-flops on a board and wire-wrapping a circuit together.[12] Even the PLDs now have an MCU core that you can cram in there for general computing needs.

In conclusion, Boolean logic is the foundation of all things digital. It is a relatively simple concept that can do some very complex things. It is clearly becoming a digital world. When was the last time you saw the latest widget marketed as the coolest new "analog" technology?

Thumb Rules

- Every significant digit you add in binary doubles value of the previous digit.
- A bit is a single piece of information with only two states, 1 or 0.
- 4 bits to a nibble and 8 bits to a byte.
- 1 is true, 0 is false.
- Always look at the truth table.
- At some point in the circuit, a signal is considered either high, 1 or low, 0; what it is depends on the thresholds of the part.
- Timing is very important when setting up more complex logic circuits.

[12] Have you noticed that the older you get, the more natural it seems to enter a state of blissful reminiscing? What could be the evolutionary benefit of that?

Microprocessor/Microcontroller Basics

This is one of the most rapidly changing fields in the electronics industry. You can purchase microcontrollers today with only 6 pins with just a few lines of memory at a cost of twenty five cents, to high-end embedded processors that just a few years ago would have been labeled supercomputers for just a few bucks. All this from the few semiconductor types we have discussed. I will not try to cover a specific processor as there are libraries of books dedicated to understanding particular micros. Instead I will try to cover some fundamental rules that can be applied in general.

Add a bunch of logic gates together, mix with some adders, instruction decoders and memory cells. Hook it all up to some input/output pins, apply a clock source, and you get a microcontroller or microprocessor.

These two devices are very similar and you will hear the names used somewhat interchangeably. Generally, however, the microcontroller is more all-inclusive, with all the elements needed to operate included in one piece of silicon, typically making them a little (but not much) more specialized. The microprocessor by contrast needs external memory and interface devices to operate. This makes it more open-ended, allowing memory upgrades without changing the chip, for example. As this area of technology has progressed, the line of distinction between these two components has blurred considerably. Hence, much of the design philosophy needed to make the most of these devices is the same.

What's Inside a Micro?

It may seem like magic, but all that is inside of a microcontroller is a whole lot of transistors. The transistors form gates and the gates form logic machines. Let's go over some of the parts that are in a micro.

Instruction Memory

I would call this ROM or read-only memory, but these days there are a lot of micros that can write to their own instruction memory. This can be programmable memory, hard coded, flash, or even an external chip that the core reads to get its instructions. The instructions are stored as digital bits, 1s and 0s that form bytes that represent instructions.

Databus

This is the backbone of the micro, the internal connections that allow different parts of the micro to connect internally. Virtually everything that happens inside a micro will at some point move through the databus.

Instruction Decoder

This is one of those logic type circuits. It interprets the instruction that is presented and sets the corresponding tasks into motion.

Registers

These are places to store data; they are literally the memory cells that we discussed earlier. This is the RAM inside the micro. It is the scratch pad for manipulating data. It can also be accessed on an external chip in some cases.

Accumulator

This is a type of special register that usually connects directly to the arithmetic logic unit (ALU). When a math function is performed on a piece of data in the accumulator, the answer is left in the accumulator, hence it accumulates the data. On a lot of the newer micros, nearly any register can be used in a similar manner.

ALU

The arithmetic logic unit, this is a part that can perform various mathematical and logic operations on a piece of data.

Program Counter

This counter keeps track of where the micro is in its program. If each piece of memory were a sheet of paper with a number on it, the program counter is the part that keeps track of the number on the sheets. It indexes or addresses which sheet it is on.

Timer Counters

These are useful when creating a structure for your code to operate in. Sometimes called RTCC (real-time clock counters), they are counters that usually can run from an independent source. They will "tick" at whatever interval you set them up to without any other intervention. Sometimes they can be hooked up to external clock sources and inputs. Usually they can be set to generate an interrupt at a preset time.

Interrupt

Not exactly a specific hardware component in a micro, the interrupt is so important that it warrants mention. An interrupt is a monitoring circuit that, if triggered, makes the micro stop what it is doing and execute a piece of code associated with the interrupt. These signals can be generated by internal conditions or external inputs. Typically only certain pins can drive interrupts.

Mnemonics and Assemblers

We humans, unlike machines, have a tough time remembering endless streams of binary data. Even trying to remember all the hex codes for a micro is very difficult. For this reason mnemonics were invented. Mnemonics are nothing more than a code word for the actual binary data stored in the instruction memory.

An assembler takes these code words and changes them to the actual data creating a file that is then copied into the instruction memory.

This differs somewhat from compilers used to compile code that you write for a computer. The compiler takes a code language such as C, for example, and creates code that will run on the computer. However, the compiler will handle tasks such as addressing memory without any need for you to worry about it, unlike an assembler. This is why they are called higher-level languages. Assembly language, as it is called, works directly with the hardware that the chip is hooked up to.

There are a lot of micros these days that have C assemblers, allowing you to use a language you are familiar with to write code for your micro. However, use caution with this approach. It is possible to lose a lot of efficiency this way. I know of one case where a micro with 4K of memory was being used to control an electric toothbrush. The developers coding in C kept coming back for micros with more memory because they couldn't get their code to fit. Once it was written in assembly, the whole thing took about 500 bytes of code. This is an extreme case. I'm sure there are much more efficient designs out there using C. Just be sure you have an idea of what your code is turning into.

Structure

The various ways you can structure your code are as infinite as numbers themselves. There are some basic methodologies that I wish I had been taught before someone handed me a chip and an application note in the lab.

Most microcontrollers only do one thing at a time. Granted they can do things very fast so as to appear to be multitasking, but the fact is at each specific instruction only one thing is being accomplished. What this means is that timing structure can have a huge effect on the efficiency of a design.

Consider this simple problem. You have a design where you need to look at an input pin once per second. One way of doing this is as follows (note the use of darrencode, a powerful and intuitive coding tool. Too bad it doesn't run on any known micro ☹):

```
Initialization
Clear counters
Setup I/O

Sense input
    Read pin
    Store reading

Delay loop
    Do nothing for 1 microsecond
    Jump to Delay loop 100000 times
Delay done

    Jump to Sense input
```

There is a slight problem with this method that you may have already noticed. The processor spent the whole time waiting for the next input, doing nothing. This is fine if you don't need the chip to do anything else. However, if you want to get the most out of your micro, you need to find a way to do something else while you wait and come back to the input at the right time. The best way to do this is with timing interrupts.

An interrupt is just what it says. Imagine you have an assistant that you have told to watch the clock and remind you right before 5:00 p.m. that you need to go to that important meeting. You are hard at work when your assistant walks in and *interrupts* you to let you know it is time to go. Now if you are as punctual as one of these chips, you drop whatever you are doing and go take care of business, coming back to your task at hand after you have taken care of the interruption. In micro terms this is known as servicing the interrupt.

Most micros have a timer that runs off the main clock, which can be set to trigger an interrupt every so often. Let's solve the previous problem using interrupt timing and see how it looks.

```
Initialization
Setup Timer Interrupt to trigger every 1 microsecond
Clear counters
Setup I/O

Main loop

    Calculate really fast stuff

Tenth second loop
    Check tenth second flag
    Jump to End tenth if not set

    Do more tasks
    Call some routines
End tenth

Second loop
    Check second flag
    Jump to End second if not set

    Read pin
    Store reading
End second
    Jump to Main loop
```

```
Timer Interrupt
    Increment microsecond counter
    If microsecond count equals 10000
        set tenth second flag
        increment tenth counter
        clear microsecond count
    Else clear microsecond flag

    If tenth count equals 10
        set second flag
        clear tenth count
end interrupt
```

One thing to note is that you don't want to put a lot of stuff to do inside the interrupt. If you put too much in there you can have a problem known as *overflow*, where you are getting interrupted so much that you never get anything done. I'm sure you have had a boss or two that helped you understand exactly how that feels. In the darrencode example above, the only thing that happens in the interrupt is incrementing counters and setting flags. Everything that needs to happen on a timed base is done in the main loop whenever the corresponding flag is set.

The cool thing is now we have a structure that can read the input when you need it to and still have time to do other things, such as figure out what that input means and what needs to be done about it. This structure is a rudimentary operating system. In my case, I like to call it darrenOS. Feel free to insert your name in front of a capital O and S for the timed code you create on your next micro. (Insert your name here)OS is free domain, and I promise you won't get any spyware using it!

The biggest downside to this type of structure, in my opinion, is the added complexity to understand how it works. The first example is straightforward, but as you step through the second example, you might notice it is a bit harder to follow. This can lead to bugs in your code simply because of the increased difficulty in following the logic of your design. There is nothing wrong with the first example if you don't need your micro to be doing anything else. However, the timing structure in the second design is ultimately much more flexible and powerful. The trade-off here is simplicity and limited code execution for complexity and the ability to get more out of your micro.

Some of you out there with some coding experience might now be saying, "Why not just run the input pin you need to check into an interrupt directly and look at it only when it changes?" That is a good question. There are times when this interrupt-driven I/O approach is clearly warranted, such as when extreme speed in response to this input is needed. However, in any given micro, you only have a few interrupts available. If you did that on every I/O pin, you would soon run out of interrupts. Another benefit of this structure is that it will tend to ignore noise or signal bounce that sometimes happens on input pins that are connected to the outside world.

Some Math Routines

It's not too hard to write a routine to multiply or divide. It can be difficult, however, to write *good* multiply and divide routines. Some of the characteristics of good routines are: short, concise, and consistent use as little memory as possible.

I've talked with students and other professionals and asked them how they would write multiply and divide routines. Remember you only get to use add, subtract, and other basic programming commands in these small micros that are so cost-effective. The most common approach that engineers come up with is the same method that I first came up with when I tackled the problem. Here is an example:

We want to multiply two numbers A * B:

1. Result = 0
2. If (B = 0) Then Exit
3. Result = Result + A
4. B = B – 1
5. If (B = 0) Then Exit Else GOTO 3

We want to divide two numbers A/B:

1. Result = 0
2. Remainder = A
3. If (B < A) Then Exit
4. Remainder = Remainder – B
5. Result = Result + 1
6. GOTO 3

These routines will work and they have some advantages: you use very little RAM or code space, and they are very straightforward and easy to follow. However, there is one significant disadvantage to them—these routines could take a long time to execute.

The multiplication routine, for example, would execute quickly if B = 3, but if B = 5,000, then the routine would take much, much longer. The divide routine runs into the same problem as the ratio of A to B becomes very large. Anyone who spends their days trying to squeeze performance out of the bits and bytes world, know that this is a no-no. Routines like this would cause you to spend all your time trying to find out why the chip resets, because of watchdog timers expiring when a big number gets processed.

Fortunately, there is a better way. I was shown the following methods and I pass them on to you as useful tools. It isn't a great secret; you just need to get out of that old mundane base-10 world and think like a computer.

The binary world has one reoccurring advantage: when you shift numbers to the left once, you multiply that number by 2. If you shift numbers right once, you divide by two. Not too hard, right? After all we've followed a similar rule since we were little in our decimal world. Shift one digit to the left and we multiply by ten, shift one digit to the right and we divide by ten.

Using this simple rule with addition and subtraction, we can write multiply and divide routines that are accurate, expandable, use very little code or RAM, and take approximately the same amount of cycles no matter what the numbers are. My examples below will be byte-sized for simplicity, but the same pattern can be used on operands of any size. You just need the register space available to expand on this.

Multiplication

Let's start with two numbers A * B. For this example, I will say that A = 11 and B = 5.

In binary, A = 00001011 and B = 0000101.

When multiplying two-byte sized numbers, I know that the result can always be expressed in two bytes. Therefore, RESULT is word sized, and TEMP is word sized. COUNT needs only to be one byte.

1. RESULT = 0 ;This is where the answer will end up
2. TEMP = A ;necessary to have a word-sized equivalent for shifting
3. COUNT = 8 ;This is because we are multiplying by an 8-bit number
4. Shift B right through carry ;Find out if the lowest bit is 1.
5. If (carry = 1) then RESULT = RESULT + TEMP
6. TEMP = TEMP + TEMP ;Multiply TEMP * 2 to set up for next loop
7. COUNT = COUNT – 1
8. If (COUNTER = 0) then exit else GOTO 4

Look at the mechanics of this. As I rotate or shift B through carry each time, I am basically moving left in B each time through the loop and deciding whether B has a 1 or a 0 in that location. (Remember, moving left is multiplying by two.) At the same time, I am shifting TEMP left each time since the binary digit I am checking in B is double the magnitude as the previous time through the loop. Then all that is left to do is add the TEMP value if the value of the binary digit in B is 1, or don't add it if B has a 0 in that location. By the time COUNT = 0, you have the final result in RESULT. The loop works the same no matter how large your numbers are. The subroutine has a somewhat small range of possible machine cycles that it takes and still remains compact and uses a minimal amount of RAM.

Let's look at our example problem each time it reaches step #8. Note x = Don't care.

Loop count	*RESULT*	*B*	*TEMP*	*COUNT*
1	00000000 00001011	x0000010	00000000 00010110	7
2	00000000 00001011	xx000001	00000000 00101100	6
3	00000000 00110111	xxx00000	00000000 01011000	5
4	00000000 00110111	xxxx0000	00000000 10110000	4
5	00000000 00110111	xxxxx000	00000001 01100000	3
6	00000000 00110111	xxxxxx00	00000010 11000000	2
7	00000000 00110111	xxxxxxx0	00000101 10000000	1
8	00000000 00110111	xxxxxxxx	00001011 00000000	0

Division

Now that multiplication is clear, division is just multiplication in reverse. Let's take the numbers A = 102 and B = 20 and perform A/B.

In binary: A = 01100110 B = 00010100.

Since I am dealing with integers, I know that A/B has a RESULT less than or equal to A .

Therefore RESULT is one byte, and REMAINDER is one byte. TEMP is two bytes.

1. RESULT = 0 ;This is where the answer will end up
2. REMAINDER = 0 ;This is for the remainder
3. COUNT = 8 ;This is because we are dividing by an 8-bit number
4. RESULT = RESULT + RESULT
5. Shift A left through carry
6. Shift REMAINDER left through carry
7. If REMAINDER >= B then RESULT = RESULT + 1 and REMAINDER = REMAINDER – B
8. COUNT = COUNT – 1
9. If (COUNTER = 0) then exit else GOTO 4

This might seem somewhat foreign, but it's really the same type of division that you've always known. You look at how many digits in the top part of A do I have to include before B will divide into those digits. Once I have a number, I subtract that division and then continue. Follow through the table with our example numbers and see if it becomes clear.

Let's look at our example problem each time it reaches step #8. Note x = Don't care.

Loop Count	***A***	***RESULT***	***REMAINDER***	***COUNT***
1	*1100110x*	*00000000*	*00000000*	*7*
2	*100110xx*	*00000000*	*00000001*	*6*
3	*00110xxx*	*00000000*	*00000011*	*5*
4	*0110xxxx*	*00000000*	*00000110*	*4*
5	*110xxxxx*	*00000000*	*00001100*	*3*
6	*10xxxxxx*	*00000001*	*00000101*	*2*
7	*0xxxxxxx*	*00000010*	*00001011*	*1*
8	*xxxxxxxx*	*00000101*	*00000010*	*0*

IT WORKS!

There are always several ways to do things, and I would never say to you that these are the best math routines for all situations. However, they are very flexible and easy to use. They can easily be adapted for 16 bit, 32 bit, 64 bit, or higher math and still work just as well.

The time that it takes for the math to execute depends on the size of the operands in bits, not the actual value of the operands, giving you more or less consistent time for the routine, a very desirable trait.

Get to Know Your I/O

One of the most important pages of the datasheet for any micro is the section that covers the I/O or the input and output pins. You should be able to answer some simple questions about the I/O of your micro. For example, how much current can the output source? How much can it sink?

Often I have had a problem getting a micro to work as I expected it to, pouring over the code trying to figure out what went wrong, only to find out that I didn't understand the limitations of the I/O pins. Don't ever assume all I/O is the same.

Knowing what your I/O is and how it works makes you infinitely more valuable as a programming resource. It sets apart the men from the boys[13] in the embedded programming world.

Some things you should know about the input pins:

1. What is the input impedance?
2. Is there an internal pull-up or pull-down resistor?
3. How long does a signal need to be present before it can be read?
4. How do you set it to an input state?

The last may seem like a strange question, but I once worked with a micro that had an input that was an input only when you wrote a high to the output port. If you wrote a low to the output port, it became an output. It was a kind of funky open-drain I/O combination.

1. Here are some things you should know about output pins:
2. What is the output impedance?
3. How much current can it sink?
4. How much current can it source?
5. How long will it take to change state under load?
6. How do you configure it to be an output?

Did you notice the timing questions? Timing, especially when accessing stuff like external memory, is important. You need to know how fast you can get the signal out of the micro and how long it takes the micro to see the signal. With timing problems, your design may work great on a few prototypes only to later in production manifest all sorts of odd behavior on a percentage of the production run.

To sum it up, it is very important to understand what your I/O can and can't do.

Where to Begin

Many times I have seen an engineer (myself included) work for hours, even days, on their code only to program a micro, sit back and…watch it do nothing. You wiggle some wires, check power and…still nothing. Where do you go from here?

[13] Or "women from the girls" in today's world you have to be politically correct even in your euphemisms.

Sometimes the best thing you can do is try to get the simplest of operations going—something like toggling an LED on and off every second. If you use the timing structure that we discussed above, getting an LED to flash will verify several things. You will know that your clock is going, you will know your interrupts are working, you will know your timing structure is in place. If you do not have an LED to flash, hook a meter or a scope up to an output pin and toggle that signal.

Once you have this LED that you can toggle on and off at will, you can begin adding to your code base the more and more complex routines you will need for a particular project. The moral of the story is "Don't try to get all your code functional all at once." Try to do some simple operations (so simple they are probably not even in the functional specification) first. Once you get some simple things down, the more complex stuff will come much easier. It is easier to chase down code-structure problems on a single LED than it is on a 32-bit DRAM data interface!

Thumb Rules

- Understand the main components of the micro.
- There are times when coding in a lower level language is preferable.
- Creating a timing structure is a way to get more out of your micro.
- Don't be afraid to use darrencode or darrenOS or create your own code and OS to help you better understand what is going on.
- Know your I/O.
- Start by simply toggling an LED with your code and go from there.
- Have a smart brother who thinks in binary.[14]
- Do simple things with your code first.

---

[14] The part on math routines is adapted with permission from an article my brother Robert Ashby wrote several years ago. Pretty slick, isn't it! He has a book on Cypress Psoc micros out that I highly recommend if you want to use that chip. Next to the guys who designed the part, he knows more than anyone I know about the ins and outs of that puppy! *Designer's Guide to the Cypress PSoC*, Elsevier, 2005, ISBN 0-7506-7780-5.

Input and Output

Input

Like the robot in the movie *Short Circuit*, all the circuits you will ever design will need input. Ergo, it seemed appropriate to dedicate a few lines to some common input devices and a little info about them. There are a few different ways to get this input.

One method is via an interrupt. You can hook a signal into a pin that can interrupt the micro. When it does, the micro decides what to do about it and moves on. This has the advantage of getting immediate attention from the micro.

Another way to monitor an input line is to use a method called polling. Polling works the same way those annoying telemarketers[15] do. They decide when to call you and ask for information. In the same way, the micro decides when to look at a pin, and it polls the pin for information.

A third way, becoming more and more common with even the smallest micro, is to take an analog reading. By nature this is a polling operation. You need to tell the A/D when to take a reading. There are some cases, however, where a pin can be set up as a comparator, and the output of that comparison can drive an interrupt. With that in mind, let's take a look at some common input devices.

Switches

Probably the most basic input device you will encounter is the switch. A low-impedance device when it is closed, a switch that is open is not connected to anything, making it the perfect high-impedance connection. This is important to note because if you are connecting a switch to a high-impedance port, when it is open you have a high impedance[16] connected to a high impedance. This is a sure way to get some weird results. The higher the impedance, the more easily disrupted the signal. To combat this, use a pull-up or pull-down resistor.

---

[15] This is assuming they are the micro. If you are the micro, I guess they would be an interrupt.

[16] When you see the words "high impedance," think high resistance to both DC and AC signals.

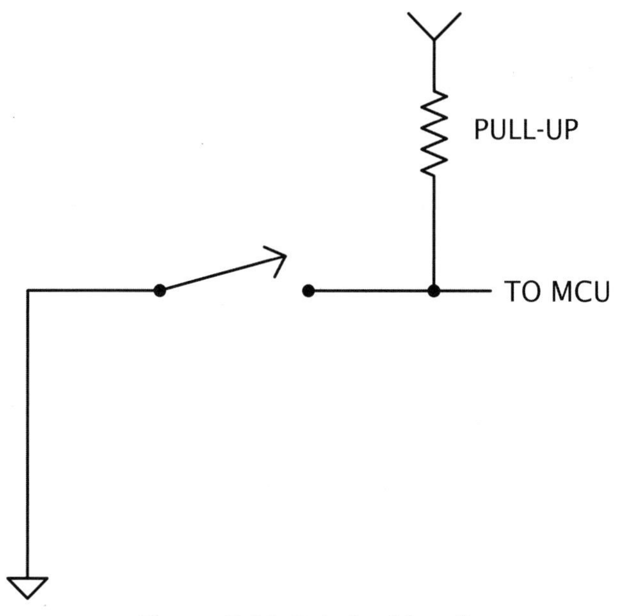

Figure 3-31 *Switch with pull-up*

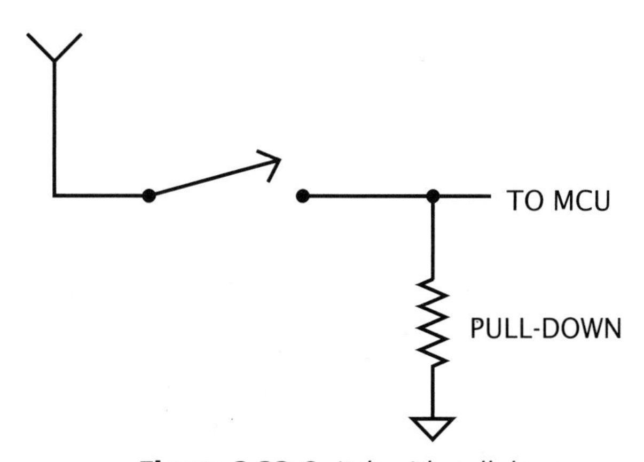

Figure 3-32 *Switch with pull-down*

Generally it is better to poll a switch input than to let it trigger an interrupt. This is because of a phenomena called switch bounce. Being mechanical in nature, a switch internally has two points that come in contact with each other. As they close, it is possible for them to bang open and shut a few times before they close all the way. The contact actually bounces a few times. The input signal to the micro looks like this.

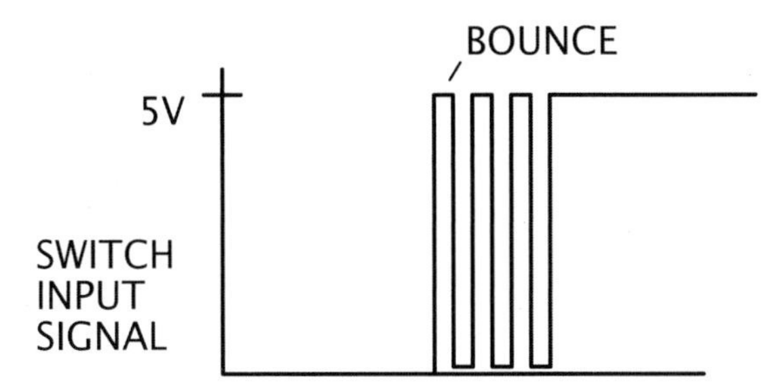

Figure 3-33 *What happens on a signal line when a switch bounces*

If this is an interrupt-driven system, you can see what might happen. Every time the signal goes high, an interrupt is tripped in the micro. When you really only wanted a single action to occur from the switch closing, you might get five or six trips of the interrupt. If you poll this line, you can determine the frequency of the bounce and essentially overlook this problem by checking less often than the frequency of the bounce. Another way to add some robustness is to require two polled signals in a row before you consider the switch to be closed. This will make it difficult for glitches or noise to be considered a valid input.

Transistors

Due to the ubiquitous usage of the transistor, it is likely you will need to interface to it as an input device at some time or another. Like the switch, the transistor is low impedance when it is on, and high impedance when it is off, necessitating the need for a pull-up or pull-down resistor. Which one depends on the type of transistor you are reading. (See the chapter on transistors.) Generally you want a pull-up for an NPN type and a pull-down for a PNP type.

Phototransistors

A cousin to the transistor is the phototransistor. This is a transistor that responds to light, often used to detect some type of movement, such as an encoder on the shaft of a motor.

You should treat it the same as a regular transistor. Note, though, that phototransistors have a gain or beta that can vary much more than a regular transistor. You will need to account for that in your design.

Another thing you should check with these transistors is their current capability. Usually they won't sink nearly as much current as the basic plain old transistor will, so don't put too much of a load on them.

Digital encoders

A cousin to the switch, a digital encoder switches lines together as you rotate the knob. Like the switch, you will need pull-up or pull-down resistors to ensure reliable readings.

Other ICs

There are a multitude of other chips out there that you can get signals from. One thing that is important when talking to other chips is timing. Often you activate the chip you are talking to with an output signal, and then you look at the data coming back. A memory chip is an example of this. You present the address on the address pins, and then grab the data from the data pins. One thing you need to consider is the time it takes for the chip to respond to this command. Every digital IC has a response time or propagation delay for it to respond to a signal. You need to make sure you wait long enough for the signal to be present before you try to get it. If there is more than one IC between you and the chip you are talking to, you need to add those delays in as well. Don't just put it together and see if it works without checking this out. It is not uncommon for a chip to be faster than the spec, so one in the lab might work fine, and yet when you get into production you will see a seemingly random failure that defies explanation.

Input Specs

Before we move on to analog inputs, there is an important thing to consider when dealing with digital inputs. Every micro has input specifications known as thresholds. These are the minimum and maximum voltages a signal must reach to be considered a high or a low. You need to make sure your signal gets above the maximum and below the minimum. If it spends any time in between, even if it seems to be working right, you could be in trouble down the road. Just remember, between those two values you can't be sure what the micro will consider the signal to be. You won't know if it is a high or low.

Potentiometers

Potentiometers (also called *pots*) are a type of variable resistor with three connections, commonly called high, wiper and low. Measuring between pins high and low you will see a resistor. The wiper is a connection that as it moves touches the aforementioned resistor at various locations. Here is a symbol of one:

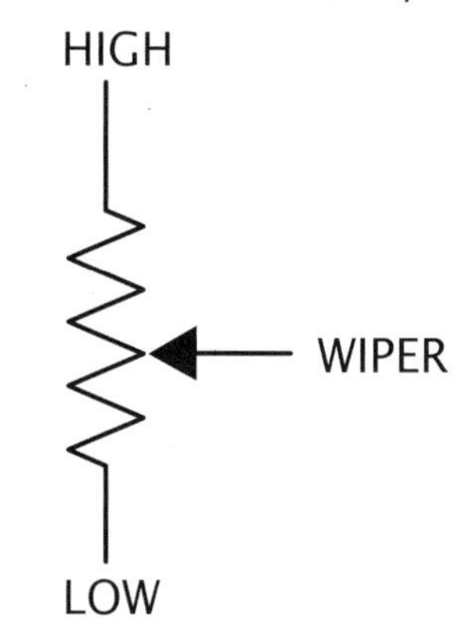

Figure 3-34 *Diagram of a potentiometer*

If you hook the input voltage to high, wiper to the output and low to ground, you have nothing more than the voltage divider that we learned about earlier. What is more convenient about the pot is this voltage divider is easily adjustable by the turn of a knob. If you tie the wiper to one end or the other as shown in Figure 13-5. You have created a variable resistor that changes as you move the knob.

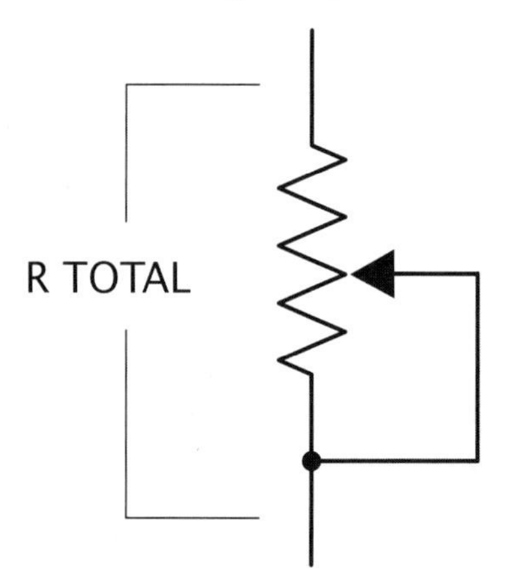

Figure 3-35 *Potentiometer made into variable resistor*

Generally pots have a large tolerance, changing by as much as ±20% in resistance high to low. However, if used in a voltage divider configuration, this variance is canceled out considerably. This is because, while the overall resistance changes, the percentage of resistance for a given position of the wiper doesn't vary nearly as much.

Analog Sensors

Thermal couples, photodiodes, pressure sensors, strain gauges, microphones and more—there are a plethora of analog sensors available. There are so many options that there is no way to cover them all, but here are some good guidelines for using various sensors.

Grounding

Where does the sensor ground go? Dealing with analog sensors requires attention to the ground as well as the power source for the sensor. Often the signal line will come right back to the chip reading it, but the ground or power leg might run past multiple ICs before getting to the corresponding pin on the chip reading the signal. This allows currents from all those other ICs to interfere with the current from the sensor.

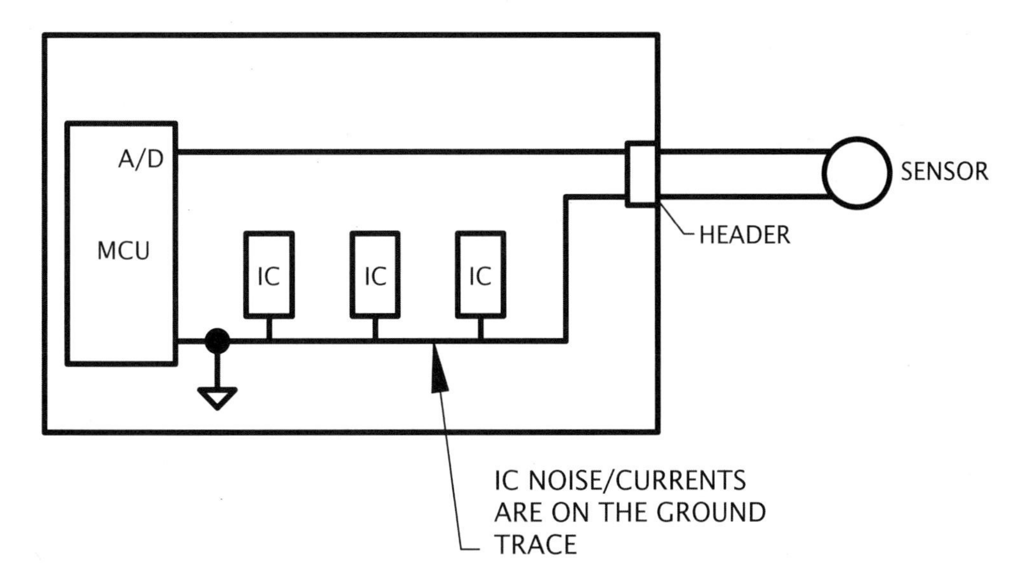

Figure 3-36 *Poor analog ground layout*

Bad, ground currents from ICs cause noise on sensor signal.

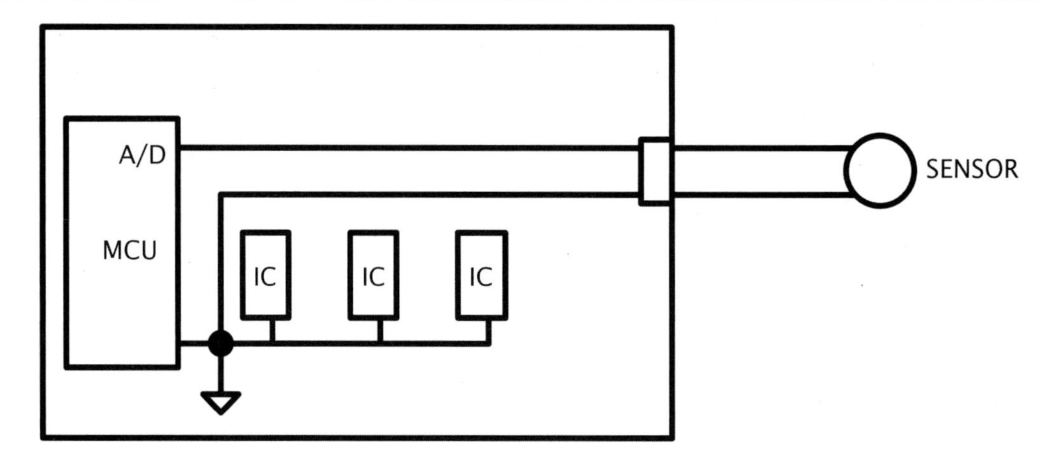

Figure 3-37 *Much better analog ground layout*

Good, traces go back to the chip, keeping the A/D reference where the A/D input is.

Sensor Impedance

What is the output impedance of your sensor? If this is too high with respect to the load it is hooked up to, it may not change the signal in the way you expect. You may need to buffer the sensor so it is not affected by loading.

Input Impedance

Most A/D converters have some type of input impedance, usually significantly lower than a digital input. A digital input is often 5M to 10M ohms of impedance, where an A/D may be 100K ohms. Get to know your input impedance, and make sure it is enough higher than the sensor output impedance so as not to be an issue. A ratio of at least 10 to 1 is a good place to start. That means if your A/D is 100K and your sensor has less than 10K output impedance, you are probably OK.

Output

There are numerous devices that you can output a signal to. We will cover a few of them here. Let's start with somc common indicators and displays. Two that are the most common these days are the LED and the LCD.

LEDs

LED stands for light-emitting diode. LEDs need current to drive them. Too little and you won't get any light, too much and they will fail, so you typically need a series resistor. How much current is needed depends on the type of LED, but 20 mA is a common normal operating current. LEDs are current-driven devices; this means the brightness depends on the amount of current flowing through them (not the voltage drop across them). This also means you can control the brightness by changing the series resistor.

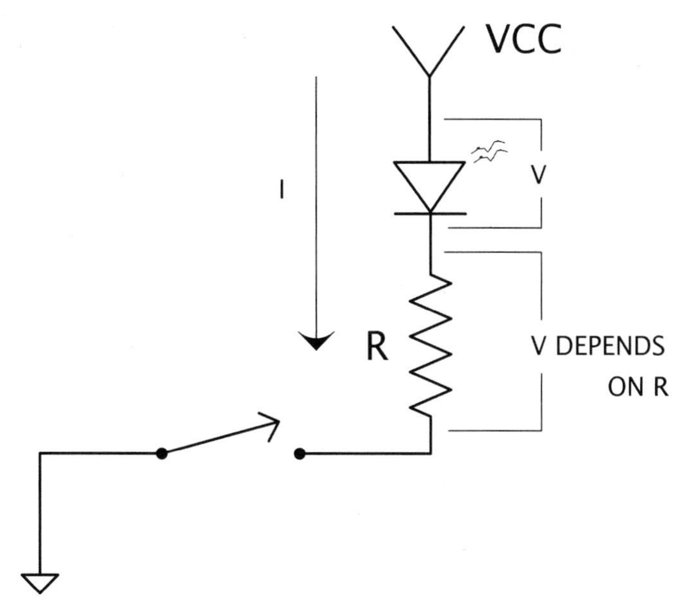

Figure 3-38 *Switch controlled LED circuit*

An important thing you should consider when driving an LED with a micro is the output capability of the chip. Does the output pin have the ability to source enough current? Can it sink enough current? There are plenty of micros out there that can sink current into a pin but can't source it. Because of this I will typically drive an LED by sinking it.

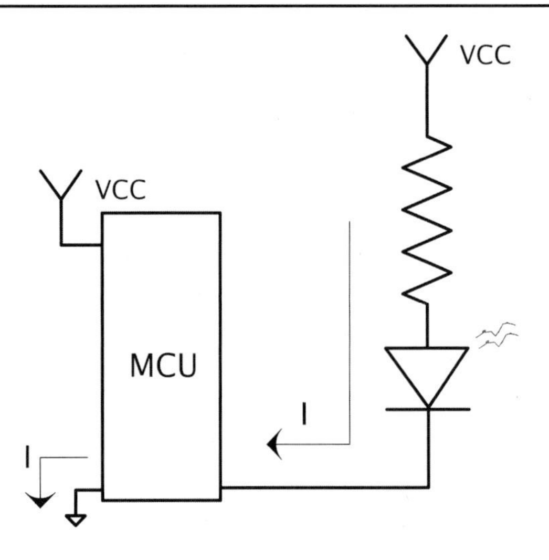

Figure 3-39 *Diode controlled by MCU*

Do you see how the current flows into the micro? You need to make sure the output pin can handle it! Also, take note that the current flows out of the ground pin on the micro and back to the source.

LEDs have a voltage drop across them just like the diode that we have already learned about. The new cool blue and white ones are quite a bit higher than the ones I was raised on. Red, green, and yellow LEDs are around 1.0 to 1.5V, while the blues can easily be 3.5V. Here is way you might consider driving one if your MCU has only 3.3V available as a supply, I wouldn't recommend it though as it has a potential problem. Do you see what it is?

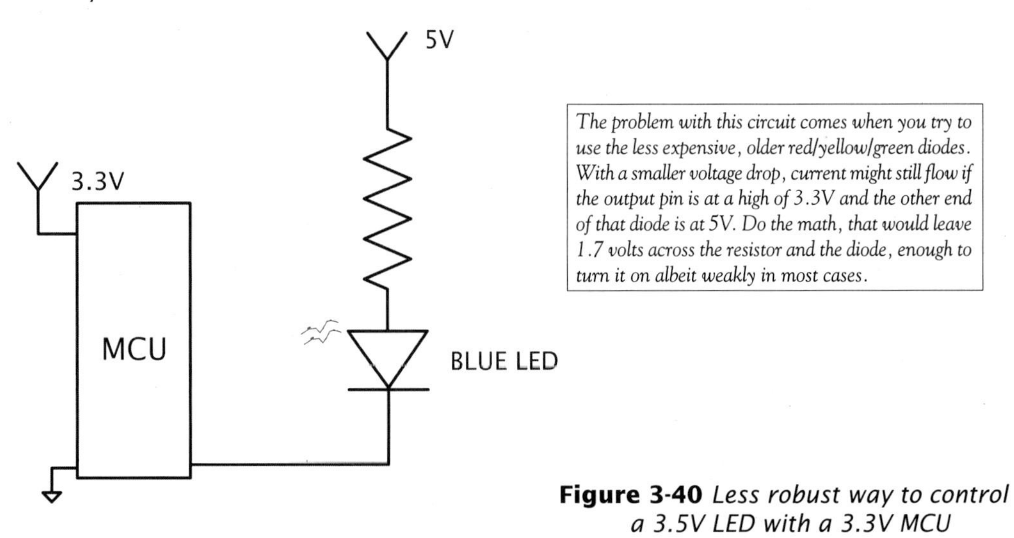

The problem with this circuit comes when you try to use the less expensive, older red/yellow/green diodes. With a smaller voltage drop, current might still flow if the output pin is at a high of 3.3V and the other end of that diode is at 5V. Do the math, that would leave 1.7 volts across the resistor and the diode, enough to turn it on albeit weakly in most cases.

Figure 3-40 *Less robust way to control a 3.5V LED with a 3.3V MCU*

Here is a better way to drive a blue LED under the same constraints.

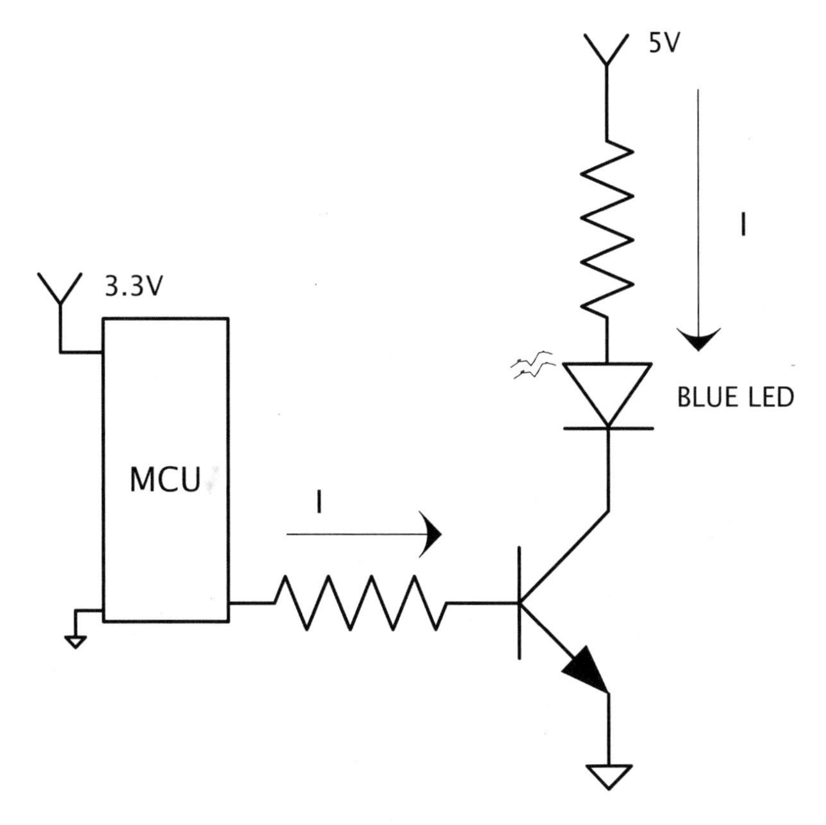

Figure 3-41 *More robust way to control a 3.5V LED with a 3.3V MCU*

The moral the LED story is, pay attention to the voltage drop needed to get current moving through it. Well enough of the pretty blinky lights, let's examine something more fluid.

LCDs

LCD stands for liquid crystal display. The liquid crystal in an LCD is a material that responds to an electric field. Applying an electric field to either side of the crystal will make the crystal molecules line up in a certain direction. If you get enough lined up, light will be blocked from passing through it.

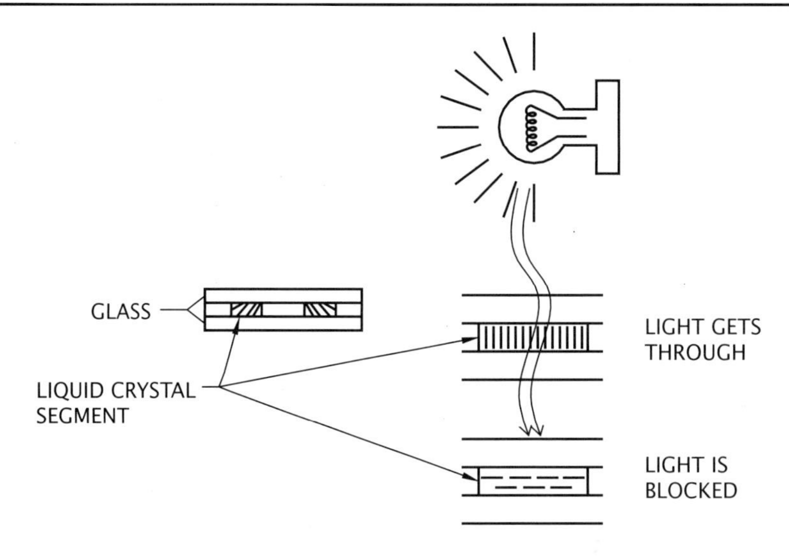

Figure 3-42 *What's inside an LCD*

If you leave an LCD biased for too long, the liquid crystal will will permanently twist and you won't be able to twist it back. It is like the crick in your neck that you get from sitting in front of the computer too long. If you don't get up and move a bit, you will tend to stay that way. While good entertainment for fellow employees, a little motion will save you the pain.

The same philosophy works with LCDs. Every so often, reverse the polarity on the LCD and all the crystals will swap direction. They still block the light, but they are all pointing the other way.

This makes driving an LCD a bit high maintenance, since you have to keep coming back to it to tell it to swap things up. It gets even more complex when you begin to multiplex the LCDs too. You need to make sure you don't leave a cumulative DC bias on one of the segments too long, etc., etc., etc.

For this reason, there are LCD driver chips. Sometimes this feature is built right into the micro, while other cases will require a separate chip. You can go it alone and make your own driver, but I don't recommend it. It is easy to mess this up, and LCD drivers are pretty cheap.

Since it is an electric field that changes the LCD, driving the LCD is a bit like driving a capacitor. Every time you switch the LCD, a little current gets by. Remember the RC circuit? But it is not much—in comparison to LEDs, it is virtually insignificant. You can get the current so low that a watch display can last for years on a battery. Remember, though, the larger the segment, the larger the cap[17], and this means more current.

Multiplexing

How do you do more than one thing at a time? Actually you don't—you do several things quickly one at a time so that it appears that you are multitasking. (Like listening to your spouse while you are watching TV. A timely nod of the head can do wonders...)

In the world of sparkies, it can be useful to multitask. One way to do this is by the art of multiplexing—that is, using fewer inputs to drive more outputs. Take a look at the example below.

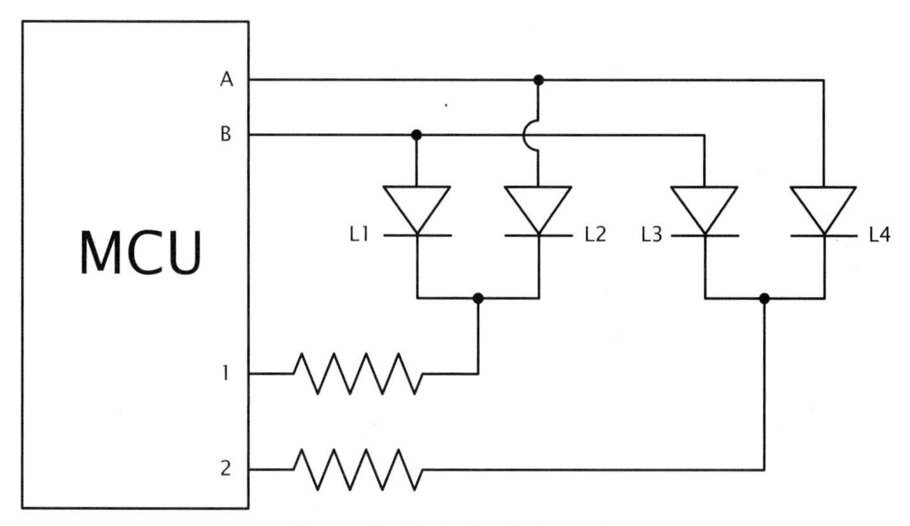

Figure 3-43 *Multiplexing LEDs*

[17] Remember that capacitance is a function of surface area.

In this case you can enable current to go through L1 and L2 by putting a low signal on pin 1 and a high signal on pin 2.

Due to the diode nature of the LED (think one-way valve, remember?), with a low on pin 1, putting a high on pin A or B will illuminate the appropriate LED. Reversing pins 1 and 2 will enable L3 and L4 to be illuminated. Repeat this process fast enough and to the human eye the LED will appear to be continuously lit. In this example we use four pins to talk to four LEDs just to keep things simple, but increase the number of LEDs in each bank and you will quickly see how fast the number of LEDs you can talk to increases as compared to the pins used. With three LEDs per bank you have five pins running six lights; with four you have six pins running eight lights and so on. If you had two banks of eight you will have ten lines controlling 16 LEDs! That is handy, especially when I/O is critical on that project where the PHB told you no, you can't have that more expensive micro with all the extra I/O. Remember, though, this slick application relies on the fact that the diodes only pass current in one direction.

Incandescent Lamp

Another indicator, this is basically a light bulb. A resistive element in a vacuum tube heats up so much it gives off light. The fact that it heats up so much should trigger the light bulb over your head, saying to yourself, "I bet that it uses a lot of current!" Which it does; it is rare that a micro has enough current capability to drive a lamp direct from a port pin.

Transistors

A great way to change the voltage (as we saw with the blue LED circuit) or to step up the output current capability of a micro. Don't forget to use a series resistor to the base; you need to limit the current as you are switching a diode to ground.

Coils

All sorts of devices have coils or inductors in them that you can send signals to. Let's take relays, for example. You may be able to drive them directly, but check the current requirements first! You will often need to use a transistor to handle the load. Also, look at the section on "catching flys" to learn about the inductive kick on a coil and what to do about it.

Thumb Rules

- Use pull-up or pull-down resistors to assert an input signal when the input device is high impedance
- Interrupt-driven inputs stop whatever the micro is doing while the line is active
- Polling inputs allow you to control when you want to look at the inputs.
- Input devices come in an infinite variety of packages and capabilities, making the datasheet on the device very important.
- You can multiplex LEDs to scare up some needed I/O.
- Transistors are a great way to change voltage levels
- Watch out for coils or inductors in devices; they may need some special consideration.

CHAPTER 4

The Real World

The real world is the place you and I live. It isn't in this book or in a simulation or even the scribbles on a schematic. All those things are representations of the real world. They help us to understand how the real world works. All the circuits we create and design at some point interface with the real world even if it is just a button to press and display to look at. It follows that we should talk a bit about some of the things we use to hook up to the big, bad world.

Bridging the Gap

If this book were written back when computers were analog, this section wouldn't even be needed. As it is, the proliferation of those pesky little digital chips gives it top billing. You need to bridge the gap between the analog and the digital at some point if you want to market your latest gadget as "way cool digital technology"! Knowing a bit about how to make the analog-to-digital leap seems like a good idea.

Analog vs. Digital

If we put analog in one corner of a boxing ring and we put digital in the other corner and then we let them duke it out, who do you think would win? In today's world, digital is all the rage, but what really sets it apart from analog? Let's find out. Let's start with analog.

What is analog? Is it just some ancient term lost in the world of today's digital engineers? No, analog basically means a continuously variable signal. It means the item being measured can be chopped up into infinite little pieces over time. Say, for example, a signal changes from A to B over a 1-second interval. If you look at it before one second is over it will likely be somewhere between A and B. It is a continuous variable. No matter how small you slice up the time segments, there is still a signal with information there. The world as we perceive it is analog in nature. Colors blend infinitely from one end of the spectrum into the other. The sound as a car races by on the street is heard as a continuously increasing and then decreasing volume level. As you drive a car you continuously change speed in response to the traffic and environment around you. The world around you is analog.

So what is digital then? My computer is, you say. Yes, this is true, but let's get a little more basic, though. Hold up one of the digits on your hand. (That is your finger, in case you were wondering.) Now put it down, now put it up again. This is digial. It is either there, or it is not. I don't know if digit (as in finger[1]) is where the term for digital came from, but it helps me remember what it means. So the simplest form of digital is two states. It's either there or not.

Let's get a little deeper. What about the time it takes to change state? What if we look at our digital finger as it moves from all the way down to all the way up? If you look at it carefully, you see that a digital signal is really analog in nature. This is true. As one of my engineer friends is fond of saying, "There is no such thing as digital, really—just funny-lookin' analog." So digital is really just a mode of perception. You look at something in a specifically determined time frame and define if it is there or if it is not. Digital is a predetermined definition of analog levels.

If digital is really analog in disguise, why even bother with it? Early on it was discovered that digital signals worked well in communication. Remember the telegraph? It used a digital dot/dash series to represent a letter. Why does it work well? Let's look at our digital finger signal example again. At a distance, it is obvious to the observer whether your finger is up or down. In fact this sort of signal is used on the freeway

[1] OK, so depending on the finger you pick for your personal example, you will either laugh or be offended. Either way you don't want to let anyone see you give yourself the "digit" as you read this in your cubicle. So I suggest the use of your index finger in this example.

every day! All kidding aside, you can avoid communication errors by using digital signals for communication.

So what is the drawback to using digital signals? The telegraph didn't last long. It was quickly replaced by analog forms of communication. The reason for this has to do with *bandwidth* (bandwidth is a measure of the amount of information a signal can carry). The analog signal can carry vast amounts of information. It can, in fact, have an infinite number of levels for a given signal range.

Back to the finger example: If you have a good telescope and can focus in on the finger, you can easily see the varying levels that the finger can represent. The same thing can be accomplished without a telescope if you have a very large finger. This implies that analog signals can represent large amounts of information much easier than digital signals can. To do this, though, it's like Tim the tool man says, "You just need MORE POWER! (grunt, grunt)." If you can't get more power out of the signal, then noise, or other unwanted information, can easily disrupt the signal. This is what happens when you get too far away from your favorite radio station and it starts to sound fuzzy. Sometimes you can give the receiver "more power" with better filters and components. But, overall, signal integrity is one of the struggles with analog systems.

To get a digital signal to move a lot of information, it has to work fast. When people wanted to hear each other talk it was much easier to use analog signals. The digital technology of the time just couldn't work fast enough to represent all the complexities of the audio information. Thus, communication efforts focused on analog encoding and decoding of information for many years. However, digital was being used in another domain entirely, in the application of Boolean logic.

Digital signals could be used to represent Boolean statements, one level indicating "true" and the other indicating "false." The computer was born. Things like if *this* is true then do *that* could now be executed by machines. Boolean logic is based on a digital representation of the world. Don't think that there are only digital computers though. For a while there were many analog computers in use to handle computations involving large amounts of information. Digital processing speeds eventually increased enough to take over these applications.

So We Have Analog

The upsides are that it can represent lots of information, and the world around us can easily be represented by analog signals. The downsides are that it takes more power in either the transmitter or receiver to resolve the analog signals, and small analog signals can be easily disrupted by outside influences.

Then There Is Digital

The pros are low power transmission, and the ability to represent logic statements. The cons are information limits (low bandwidth) requiring it to work fast to process large pieces of information and the fact that the world around us is analog, not digital in nature.

The Best of Both Worlds

Wouldn't it be great to have the best of both worlds? That's what engineers thought, so they coined a couple of acronyms to get the process started, ADC and DAC—the analog-to-digital converter, and the digital-to-analog converter. Let's find out what these are.

A-to-D and Back Again

What is A-to-D conversion (or ADC)? Is it a religious experience? Is it the opposite of D-to-A (or DAC) conversion? A to D is all about taking the real world and making it into ones and zeros so that digital technology can manipulate it. You can reasonably say that D to A reverses the process. Here we will explore what this A to D to A is and what it is good for.

First, A Is For Analog

An analog signal is converted by chopping it up into chunks at predetermined time intervals. (This chopping is called the sample rate. The faster the sample rate, the higher the frequency that can be digitized.) Then the signal is measured at that point in time and assigned a digital value. Digital signals (often represented as 1 or 0) can be crammed together to indicate different levels of analog. One digit can indicate two levels. If you use a binary numbering system, the more digits you use, the number of levels goes up by 2 raised to the power of the number of digits. Four digits give you 16 levels (2^4). Eight digits gives you 256 levels (2^8). And so on. One common way of determining the level of a signal is to use a comparator.

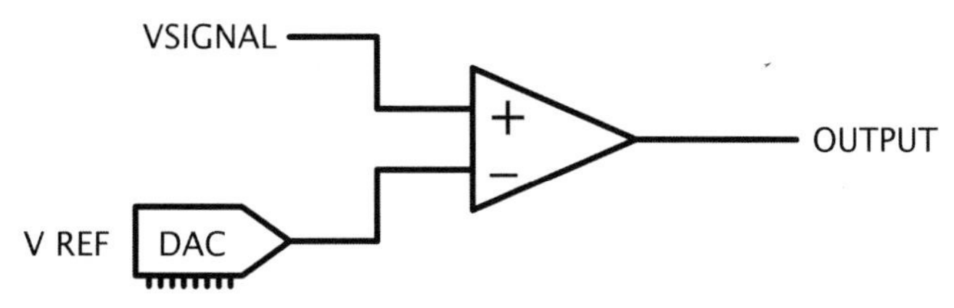

Figure 4-1 *Comparator-driven ADC*

Remember this application? In this case, the signal is compared to a reference voltage. You increase the reference voltage from min to max. When the signal is larger than the reference voltage, the op-amp comparator will output a high or a '1'. When the reference voltage is the larger of the two, the output will be low, or a '0'. If the circuit knows the value of V_{ref} at the time the output changes state, this is when V_{ref} is approximately equal to V_{signal}. I say approximately, because there is always a question of resolution. For more on this, read on.

D Is for Digital

Now we have a digital signal. We can do lots of fun things with it. We can transmit it, and manipulate it without worrying about signal loss. But what is next? Say we convert guitar music into digital format to add some neat sound effects. You can't just send the digital data back out to be heard. It must be converted back to analog. Why? Because there are certain things we perceive well in an analog format. If you don't believe me, take a look at your speedometer, I'll bet it is an analog gauge. (There are some things we like to see digitally but usually that's so we don't have to deal with infinite increments.) To convert a digital signal back to analog, the circuit has to simulate the analog signal it represents. This always requires some kind of filtering. There are many ways to convert digital to analog. One of my favorites is by pulse width modulation (PWM). In a PWM circuit, the output of the device switches on and off at a given frequency. The percent of time it is on versus off is the amount of analog signal it represents. This percentage is called the duty cycle.

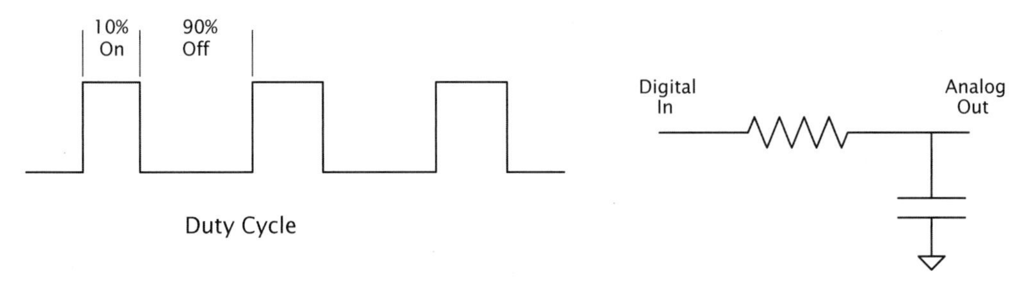

Figure 4-2 *Duty cycle-controlled analog output*

The digital PWM is fed into a low-pass filter that removes the switching frequency of the signal, essentially leaving an analog signal. The number of levels that this signal can represent depends on the resolution of the PWM signal. This means the capability of the PWM to be switched on and off at varying duty cycles. For example, a PWM that could switch on and off in increments of 5% duty cycle would have less resolution than a circuit that can handle increments of 1% duty cycle. This means digital signals can only represent discrete levels of analog signal. These levels are the resolution of the signal.

Why is resolution so important? We stated above in the comparator example that the circuit knows what level V_{ref} is at. How does it know that? It must generate it somehow. It does so with some type of DAC process. And the resolution of that DAC process will determine the resolution of the ADC process.

So there we are, we went from analog to digital and right back to analog again. It really is a circle. Let's look at some examples to see this in action.

It Takes a Little D to A to Get a Little A to D

A while ago I was explaining my thoughts on the world being analog in nature to a fellow engineer. He emailed me the following response:

> *I would like to provide counterpoint to your assertion that "the world as we perceive it is analog in nature." I think that there are as many, if not more, natural digital perceptions as there are analog. Some samples: alive or dead, night or day, open or closed, wet or dry, flora or fauna, dominant or submissive, predator or prey, hungry or full, coarse or smooth, hot or cold,*

> *fuzzy or sharp, open or closed, single or multi, camouflage or warning, flat or mountainous, forest or desert, stormy or clear, noise or silence, blind or seeing, male or female, feast or famine, survive or die, on or off and so on. Granted, things like warm, breezy, sunsets, and omnivorous are there, but for the most part, I think our nature perceptions are digital.*[2]

In many ways he is correct as is so eloquently stated, but he refers to perception. We place the analog information from the world into "digital buckets." I think the reason we do this is to facilitate decision making, to limit the store of information, and to ease communication. We impose a digital perception when it makes sense to do so. A better phrase may have been "The world is analog in nature upon which we impose our digital perceptions." With that in mind, let's look at some more of the nuts and bolts of A-to-D conversion.

ADC stands for analog-to-digital conversion. How do you do this? In many cases, you begin with a DAC (digital-to-analog converter) and a simple comparator. Remember the comparator circuit from previous articles that used an op-amp? The op-amp would output a high or low signal depending on whether one input was above or below the other. This is a great time to use a comparator, as digital circuits like obvious signals such as high and low. So here is the basic process: you convert a digital number to a known analog level, compare that to an analog signal, and if it is close to the same value (here is where resolution counts), the digital number you output represents the analog value. Refamiliarize yourself with this diagram.

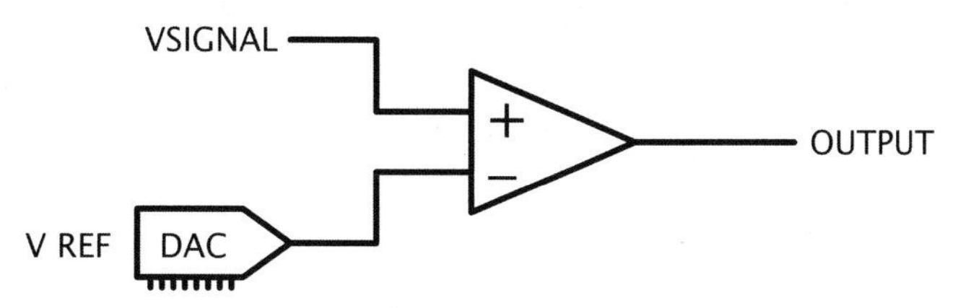

Figure 4-3 *Comparator-driven ADC*

[2] A good friend of mine by the name of Michael Angeli. I've always liked his writing style, maybe someday I could get him to collaborate on something with me.

Let's do an example. You are converting an analog signal with the actual value of 4.5. You try outputting a 1 on your DAC. The comparator says higher. You then try outputting a 2. The comparator says higher You try a 3, then a 4. Guess what the comparator says each time. That's right, It says "higher." So what do you try next? Of course, you try a 5. Then the comparator says "lower." Now your circuit knows that the value is between 4 and 5. It makes a guess and picks one of the two. The more resolution (resolution means smaller step size) that your DAC has, the closer you can estimate the value of the analog signal.

There is a better and faster way than just sweeping across all the values in the range. Make your first output on the DAC equal to 1/2 the range. In this case it would be 2.5. Then you look at what the comparator says and make a logic decision (digital is good for this sort of thing). If the comparator says 'higher," you can eliminate everything below 2.5. So you make your next output equal to half of the remaining range, which, in this case, is 3.75. Look at the comparator again, and eliminate some more possibilities, then output half of the remaining range. Repeat this process until you are out of resolution and you will have an approximation of the analog signal. This is a very fast way of approximating an analog signal known as successive approximation. It is often used when high-speed analog-to-digital conversion is needed.

Did you notice that I often use the word "approximation" as the A-to-D process takes place? This is because a digital signal can never truly equal an analog signal; it must always draw the line somewhere. Do not forget that "digital" means there are discrete steps involved. Analog has, by definition, infinite increments. Now that you have the basic idea behind the A-to-D conversion process, let's look at some examples of DAC circuits that can be used.

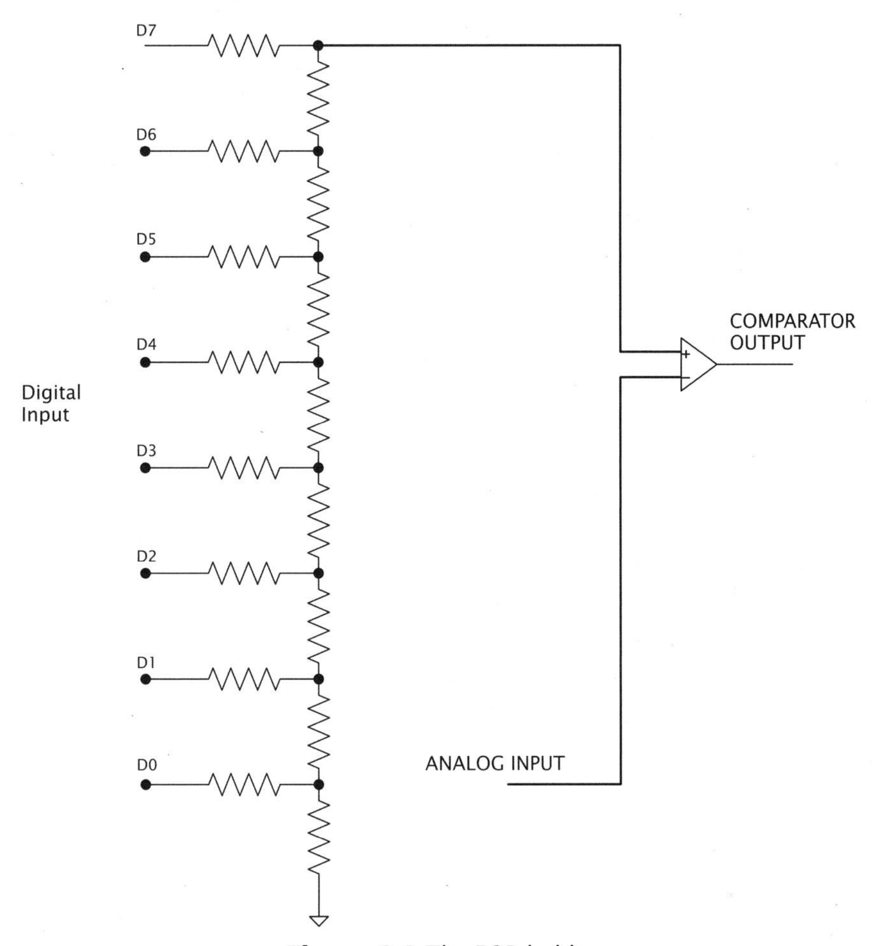

Figure 4-4 *The R2R ladder*

This is a neat way to get a digital voltage level, and you can get the R2R ladder in a nice compact package. You must take care not to hook it up to any low-impedance devices without buffering, as its output level can be easily affected by external loads.

How does the ladder work? A digital byte is output to the ladder, which changes the voltage level to the input of the comparator. You should note that the MSB (most significant bit) has the most effect on the output. The LSB (least significant bit) affects the output the least. This works very well with the approximation method described above. You simply load the DAC value you want on the resistors and look at the output signal. It is very fast. The biggest downside is that it uses a lot of output pins. (The output pins must be able to sink or source sufficient current to work correctly.) One caution: make sure your processor can handle the output load of the ladder. The Zilog processor I used in one application worked fine and even has an onboard comparator for the ADC process, but I did use every pin, leaving little room for additional signals if needed.

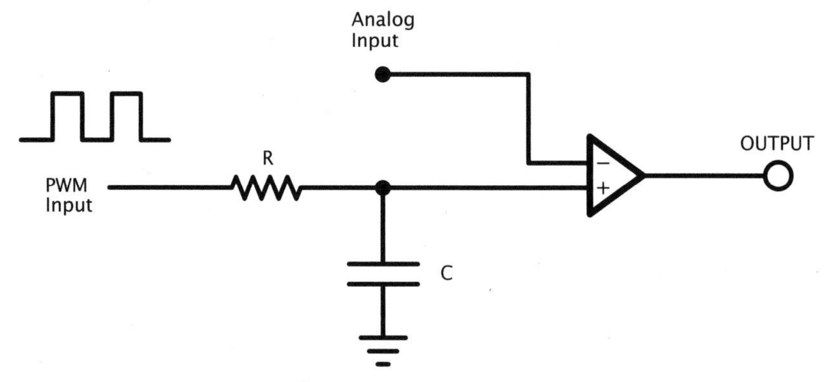

Figure 4-5 *PWM ramp*

In this circuit, the duty cycle of the PWM (pulse width modulation) signal is ramped up from 0% till it passes the value of the analog signal, as indicated by the comparator. The analog voltage is represented by the percent of the PWM signal when the comparator changes state. The RC filter must turn the PWM into basically an analog level. This means the PWM must switch significantly faster than the speed of the signal you are trying to digitize.

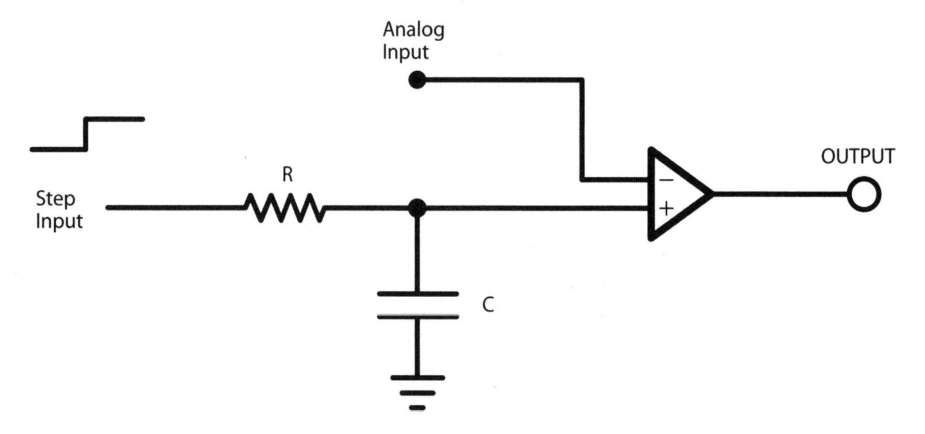

Figure 4-6 *RC Charge time*

This method relies on the transient response of the basic RC circuit. The step input causes the input to the comparator to increase according to the response time of the RC circuit. The output of the RC circuit is equal to $V_i(1 - e^{-(t/\tau)})$. So if you know the value of tau, which is R*C, then you can calculate the voltage based on the time it takes to pass the input. This can be tedious to calculate in some micros, but often high accuracy is not needed, and a look-up table of the values can be implemented. In many cases, a lower resistance discharge path is added to this circuit to assure the output of the RC circuit begins at zero. The downsides to this circuit are the curve calculations, but the first three tau of the signal are a fairly linear approximation, which, depending on the application, may be good enough. (Review the connection between electronics and hand grenades way back at the beginning the book to see when things are "good enough.") If your task isn't too demanding and you don't get too close to the upper rail, you can just count time and toss out that complex calculation, making this a quick, cheap and dirty ADC.

So there you have three easy ways to get a digital approximation of an analog signal. All of these circuits are perfectly fine to use as DACs only.

One last thought—these days a built-in A/D converter is a more and more common feature on a microcontroller. However, they nearly all work on the principle of using a DAC to make an ADC. Studying this section can help you to get an idea of what is really going on in there. The more you know about how it works on the inside, the better engineer you will be!

DSP

DSP or digital signal processing refers to manipulating data that is digitized from an analog signal. In many cases, such as audio and video, the signal is converted back to analog after DSP occurs. Many books are available on DSP that are far better texts on the subject than this one. I only hope to create a bit of understanding on this topic.

One of the advantages of a DSP is the ability to change parameters of the filters on the fly. This allows engineers to create all sorts of new solutions to processing signals that are very difficult to achieve with comparable analog designs.

Typically a DSP solution is also more expensive than an analog one, so be sure you really need it. Don't slap a five-dollar DSP chip in the circuit when a twenty-five cent op-amp will do the job. That is not to say DSP doesn't have its place. Without DSP, we wouldn't have MP3s, WMAs, AC3, MP4s, WiFi and a whole other slew of acronyms to spout about! Come to think of it, DSP technology may be responsible for more acronyms than any other…

Thumb Rules

- Analog is a continuously variable signal.
- Digital is discrete steps.
- Resolution is the distance between the discrete steps.
- DAC is often used for ADC.

Making Stuff Move—The Electromechanical World

One thing that happens in the real world is moving stuff. Eliminating moving parts is a commonly sought-after goal in the world of electronics. However, I suspect that sometime in your career you will need to make things move and you will be thrust into the world of electromechanical devices. Considering that what I knew about motors when I left school could be written on the thin edge of a postage stamp, I felt a need to cover some of the basics behind motors and a few other electromechanical devices.

DC Motors

My eldest son was elated when he got a Lego Mindstorms® kit for Christmas a few years back. For those who don't know, it is a ready-made robot kit based on—you guessed it—Legos. My wife claims I was much more excited than he was. I beg to differ, but we won't go into that now. The whole point of a robot is that it moves. The Lego kit uses little DC permanent magnet motors with gears and such to get along. Since this type of motor is so popular, I thought I might explain a little about DC permanent magnet motors and how to control them.

The DC permanent magnet brush motor is probably the easiest motor to understand. It consists of just a few parts: an armature, some magnets, a case, wires and brushes. I remember as a kid making a motor out of a couple of nails, a dowel, and some wire. It looked something like this:

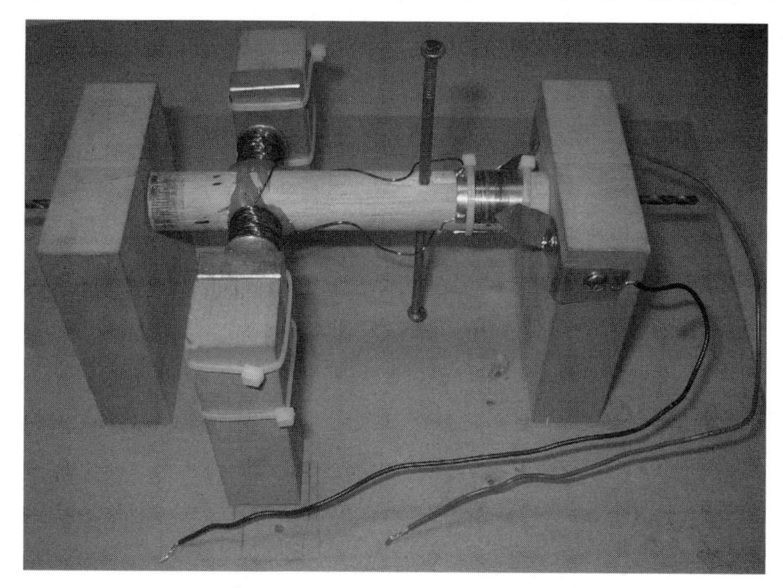

Figure 4-7 *A home-built motor*

You can make a motor by winding the wire onto the armature in a loop. The ends of the wire terminate on segments that the brushes rub on.

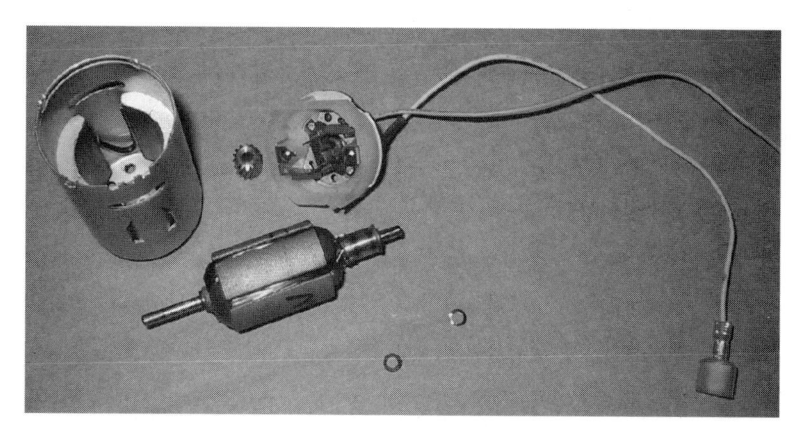

Figure 4-8 *A motor taken apart*

Permanent magnets are attached to the case in such a way as to surround the armature. The armature is supported in the case by bearings or bushings so that it can rotate freely. At its most basic, the coil of wire on the armature is nothing more than an inductor. As we learned earlier, an inductor develops a magnetic field when you pass current through it. This magnetic field is just like the one present around the permanent magnet. By controlling when the magnetic field is present around the armature, you cause the field around the wires to push or pull against the field around the magnet. The current to the armature is switched on and off (which turns the magnetic field on and off) in a sequence that causes the armature to turn. This is called commutation. In the DC PM brush motor, the brushes are the method of commutation. They switch the current through different sections of the armature as it turns.

A DC PM motor has two inputs and two outputs. You put voltage and current in and get speed and torque out. One nice thing is that the speed is proportional to the voltage and the torque is proportional to the current. Motors are a device where the physical equivalents of electric components are not only similar in nature, but actually linked in performance. Think of it this way—voltage and current together equal power. Speed and torque together also equal power. So in a motor you put electrical power in and get mechanical power out. That actually makes sense, doesn't it? The equivalent circuit looks like this:

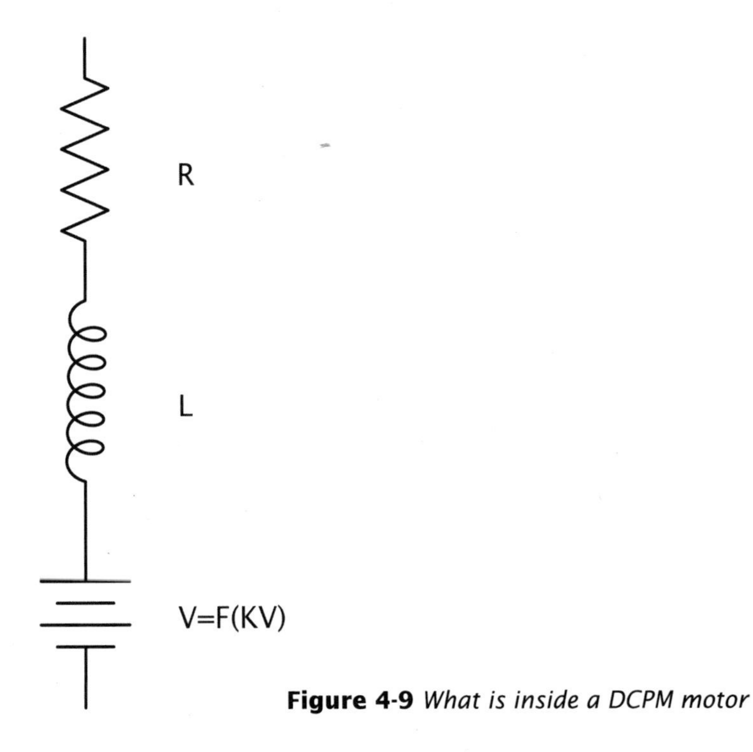

Figure 4-9 *What is inside a DCPM motor*

What do you think the resistor is doing in this circuit? Have you ever noticed a motor getting warm when it operates? This heating comes from the resistive component in the motor. Any wire short of a superconductor has resistance. The armature, being made out of wire, also has resistance. Current flowing through a resistor will create a voltage drop across said resistor, and power across that resistor turns into heat. Ohm's Law still works. The inductor creates the magnetic field that turns the armature. The battery represents what is called the back EMF, or electromotive force. If you were to spin the shaft of the motor with nothing but a voltmeter hooked up to it, you would see a voltage appear on this meter that is proportional to the speed you spin the shaft. When you apply a voltage to the motor, the shaft will spin at a speed by the same proportion. However, not all of the voltage you apply to the leads makes it to this point in the motor. Some of it is lost across the resistor. All this leads to some characteristic equations of this type of motor.

The relationship between voltage and speed is known as the voltage constant, with units of volts per Krpm. It is referred to as K_e or K_v.

$$K_v = \frac{V - IR}{Krpm} \qquad \text{Eq. 4-1}$$

V is the amount of voltage applied at the leads.
I is the current flowing through the motor.
R is the equivalent resistance of the motor.[3]
Krpm is the speed of the shaft in thousands of revolutions per minute.

The *IR* term in this equation accounts for the loss in heat in the motor. As current approaches zero, this effect disappears. This is what happened above when we hooked it up to a voltmeter and spun the shaft, reading the voltage generated. Do you see how that minimizes the error, giving you an accurate idea of the voltage constant?

[3] Note that you can get a fairly close idea of this with a simple ohmmeter turning the armature very, very slowly. (Too fast and the voltage generated will mess up the reading.) To be more precise, you need to take the resistance of the brushes and the way they contact the armature into account, a discussion that we will save for another book.

The relationship between current and torque is known as the torque constant, usually referred to as K_i, it has the units in-oz/amp

$$K_v = \frac{V - IR}{Krpm}$$ Eq. 4-2

T is torque in inch-ounces.
I is the current in amps.

These two constants are linked; changing one changes the other. In fact, if you know one, you can calculate the other with this equation.

$$K_i = \frac{T}{I}$$ Eq. 4-3

$K_t = K_v$ [Nm/A; V/rad/s]
$K_t = 9.5493 \times 10 - 3 \times K_v$ [Nm/A; V/Krpm]
$K_t = 1.3524 \times K_v$ [oz-in/A; V/Krpm]

As you can see, it turns out we are really only dealing with one constant in the motor. This constant is controlled by how many windings are on the armature and how strong the magnets are. More windings increase the voltage/torque constant, fewer decrease it. The size of the armature and the strength of the magnets also affect this constant.

We now know that the main electrical components of a motor are resistance, inductance and a voltage source. Can you extrapolate the mechanical properties? They are friction and inertia.[4] The first thing you should note is that the load, or whatever is hooked to the motor shaft, will likely be the largest contributor to these two characteristic factors, masking the effects of the armature inertia and brush or bearing friction.

Inertia will tend to make the motor take time spinning up to speed, increasing the load and current draw as you accelerate. Once at speed inertia will tend to keep the motor spinning so during deceleration you will notice the current the motor needs lessens.

---

[4] You could also have a spring-type component as we discussed way back in the beginning of the book, but it is pretty rare to find that in a DC PM motor.

Friction will create a constant load on the motor which will appear as an increase in current in our "sparky" universe. To gain further light and knowledge on all things motor, I refer the reader to the "pink book."[5]

"But what about the Lego Mindstorms?" you ask. Well, my boys and I have built quite a few projects, but my oldest son has lost interest since we couldn't seem to build a robot that would clean his room. I told him that I sincerely hope he can someday solve that particular problem, but till then it is up to him.

DC Motor Control

Given what we just learned about this type of motor, I hope it is apparent that if we want to control the speed of a DC PM motor, we should control the voltage to it. If we want to control the torque we should control the current. If this doesn't make sense, take a look at the DC motor equations once more.

Speed control

Let's start with a simple application. Say you want to spin a motor at 500 rpm. This motor has a K_v of 10 V/Krpm. Plug that into the equations we just learned and you find out you need about 5V to get this motor going at the speed you want (neglecting load for a moment). But how do you go about getting 5V to the motor? There are two different ways to approach this problem. You can use a linear methodology or a switching methodology. In both cases, you will start with a higher voltage than you want at the motor and then lower it and apply it to the motor leads. We should go over both types of systems to understand the pros and cons of each.

Linear Control

The simplest way to make a linear control is based on the voltage divider rule. Put a resistor between the power supply and the motor, and adjust the value of the resistor until you have the amount of voltage you want across the motor. Something like the following:

---

[5] *DC Motors Speed Controls Servo Systems*. I like to call it the pink motor book due to an interesting choice of color for the cover, and I highly recommend it for anyone who is working with DC motors.

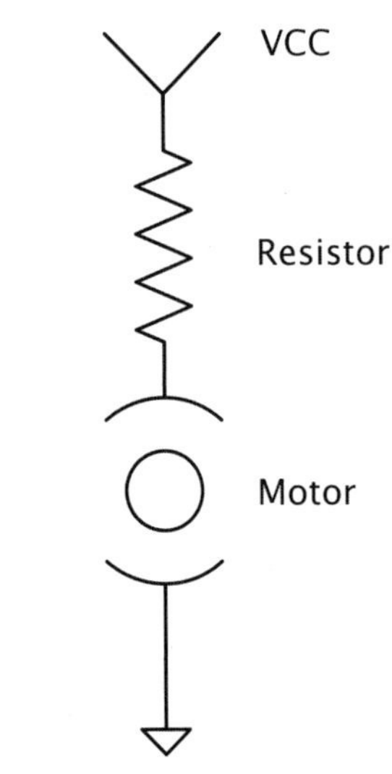

Figure 4-10 *A motor with a resistor in series*

The biggest drawback about this design is caused by the fact that the motor can be a dynamic load. As the load on the motor changes, the amount of current through the motor changes, which, following Ohm's Law, changes the voltage drop across the resistor, which changes the voltage across the motor and hence the speed of the motor is destined to change.

However, if the load is consistent, or if variability is OK, you can dial this design in and make it work fine. You should note that the resistor will heat up based on the current through it and the voltage across it. For example, if Vcc is 10V and you set the resistor value such that 5V is across the motor, this means there is 5V across the resistor. If the current drawn by the motor in this case is 1A, you will need a 5W resistor to handle the power. (Actually, any engineer worth his salt will not run the power resistor at its maximum wattage but will overate it liberally.)

In this linear control design. the resistor can be replaced by a FET or transistor or some other type of amplifier operating in linear mode, allowing the voltage to the motor to be adjusted as desired. By using feedback methods as previously learned, the variation in load can also be compensated for so you can maintain the desired voltage to the output.

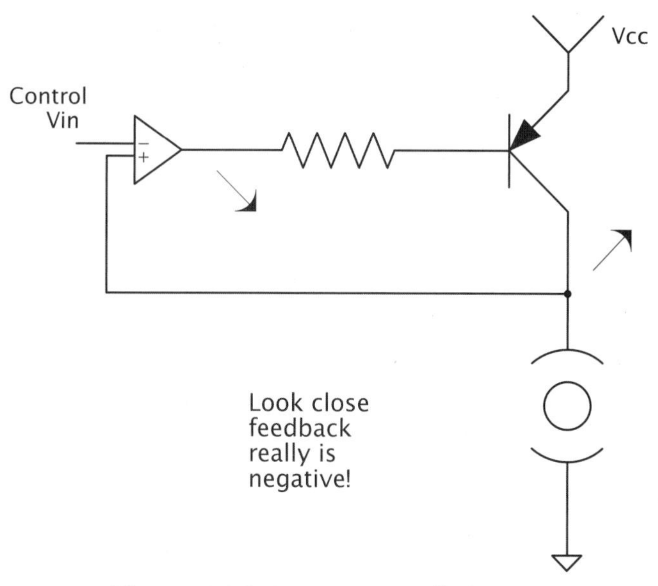

Figure 4-11 *Op-amp controlled motor*

This design has the significant advantage over the previous one of maintaining the voltage to the motor at a desired level regardless of the changes in load. This will maintain a more constant speed than the previous design, but there is still room for improvement as we will see later.

The biggest drawback to this type of design is the same as the resistor: excess power is turned into heat. One benefit, though, is as far as EMI is concerned it is a quiet design.

Switching Control

In contrast to linear control, a motor can also be controlled by switching power on and off to the motor. The similarities of switching motor control to switching power supplies are many. In many switching supply designs you will find an inductor that stores the energy when the switch is on, and discharges it to the load when the switch is off. The same thing can happen in a switching motor control. The inductor, however, is inside the motor. In the switching supply you will find a diode that directs the current from this inductor to the load. In a correctly designed switching motor control, you will find a diode that performs exactly this function.

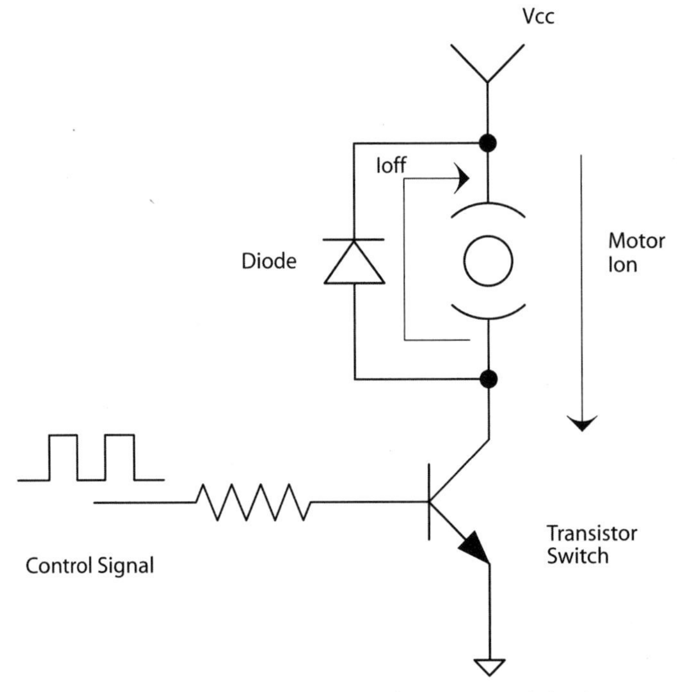

Figure 4-12 *Transistor with motor and diode*

When the switch is closed, current flows through the motor. When the switch opens, the current from the inductor goes through the freewheel diode back around through the motor. Since this current is recaptured and applied to the load, switching motor controls are more efficient than their linear counterparts.

Some important things to note: the switching frequency of this type of control needs to be fast enough for the inductor in the motor to act this way. If the switch doesn't turn back on before all the current has discharged from the inductor, you will feel the torque change in the motor (it will manifest itself as a vibration).

In a way, the inductor filters the high frequency of the switching power, reducing torque ripple and thus vibration. The most common form of control in this case is called PWM for pulse width modulation. By varying the duty cycle of the PWM, the amount of power to the motor is varied.

Switching motor controls are very prevalent these days.

Maintaining Speed

Often some sort of voltage feedback is used to maintain the output voltage of the control to a desired level. Remember that in the DC PM motor the speed of the output shaft is proportional to the voltage applied. That makes it nice for maintaining speed. However, if you flip back a few pages you will notice the *IR* component of that equation represents what are known as losses. These losses are burned up as heat across the resistance of the wire in the motor armature (plus a little in the brushes). The loss is proportional to the current *I* through the motor and the current is proportional to the load on the motor shaft.

This means that, as load varies on the motor, the amount of loss varies. This results in a change of speed. Think of it like this—the voltage that gets burned up as heat never makes it to spinning the motor shaft.

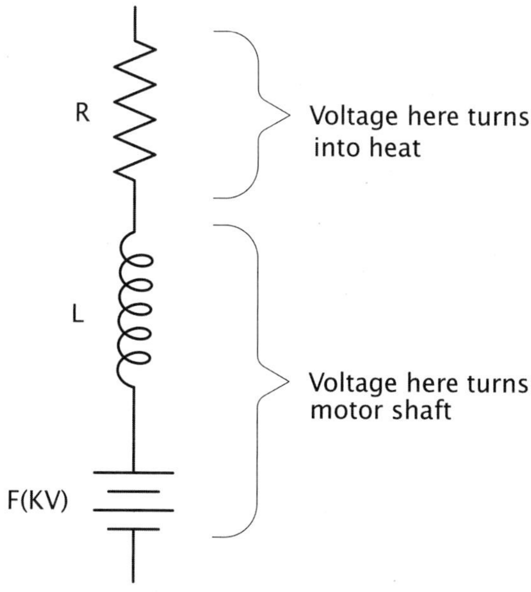

Figure 4-13 *What happens inside the motor*

There are two ways to compensate for this. One way is to use speed feedback to adjust the voltage output to the motor to maintain a constant speed. The other is to compensate for the losses themselves.

In most DC motor control designs, you will find a voltage feedback loop that does 90% of the speed control work. Then, external to that, you will find a speed feedback control loop that will compensate for the rest of the variation.

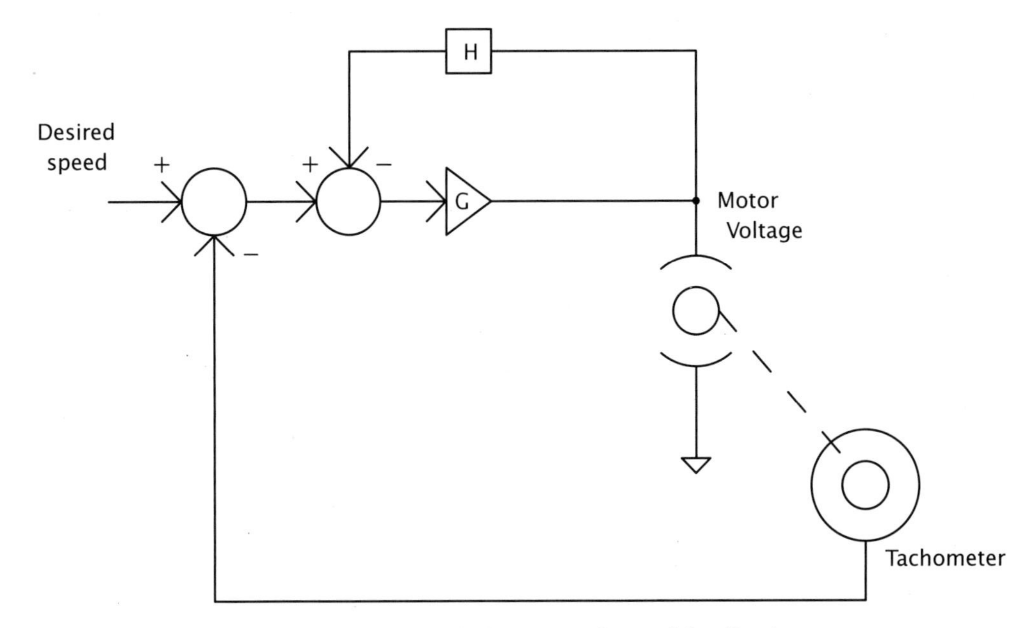

Figure 4-14 *Block diagram of speed feedback*

While it is generally a good idea to feed back the signal you want to control, there may be times you do not have that luxury, or maybe there are reasons you do not want to use speed feedback. If this is the case, you can use another speed control approximation call IR compensation. This is a method in which you monitor the load on the motor by sensing the current through the motor. The loss due to heat is proportional to this current. If you know the resistance of the motor, you can calculate how much voltage turns into heat, never making it to the output shaft. Add this much voltage to the input to the motor and you have a fairly good approximate speed control and you didn't need a tachometer!

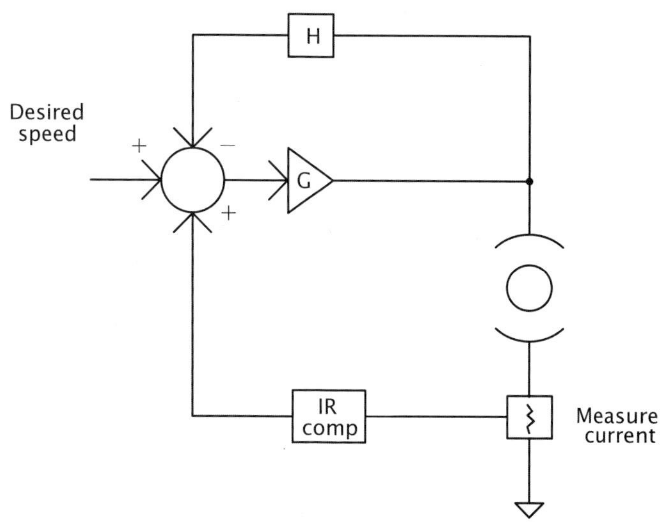

Figure 4-15 *Block diagram of IR comp feedback*

All in all, controlling a DC PM motor speed is one of the simplest motion control problems to tackle, but it is simple only relative to the other options out there. It is still a significant "chunk o' learnin" to swallow. For this reason I suggest starting here if you want to learn motion control before moving on to some of the other available motors.

Torque Control

One thing that happens when you take a motor and stop the rotor is a huge increase in current. Depending on the motor and your design, this may be more current than your control can handle. In this case you may need a current-limiting circuit.

This is a circuit that monitors the current used by the motor and at a preset level, scales back the output, limiting the current available to the motor.

In a DC PM motor, when you control the current, you control the torque. Don't believe me? Flip back a few pages and look at the torque constant equations. The units are in-oz per amp. This is a linear relationship—the more current through the motor, the more torque at the output shaft.

What all this means is that a constant current supply will create a constant torque when hooked up to a motor. This is essentially what happens when a control hits its current limit. The control goes from being a constant voltage supply to a constant current supply. This protects the motor and the control from damage.

Braking

Imagine careening down a hill on your electric scooter. "Gosh," you think to yourself, "it would be nice to use some of the energy I am wasting to slow this vehicle down. There ought to be a way to make it recharge the batteries. I'm an engineer," you think, "Why don't I design a regenerative brake?" Just such a thought has come into my head and I have been able to ignore[6] it quite effectively, until now.

Some time ago I was asked to design a motor control with a regenerative braking circuit. Having done several controls, but none with regenerative braking, I started by perusing the Internet. I don't follow Star Trek's creed to boldly go where no man has gone before on a whim. That is to say, if someone has been there already, I would sure like to know the path he or she took. Once the end of that path has been found, I will then venture into the unknown. Several hours of searching were somewhat futile. A simple and concise explanation and possibly a schematic (particularly for a PM DC motor) were all I needed. There were reams[7] of information explaining what it does but not much was there showing exactly how it was done. Alas, my effort to find the simple explanation was to no avail. Maybe it was out there somewhere, but I got sick of all the pop-ups.

As you may have guessed by now, I take such a lack as a personal affront that I must correct. The following is what I have pieced together in my own mind, distilled down to my level of intelligence (the longer I spend in management, the lower this level seems to be), then ousted to my readers in a form I hope is easy to understand. After I looked at the best idea since raw toast and the nice read about the Honda Insight's regenerative brake, this is what came out.

---

[6] I find it very easy to ignore such thoughts when I am playing Nintendo. In fact back in my college days, I had to redo an entire quarter of school (except for one class that I passed due to a very persuasive paper on said topic) due to a severe Nintendo addiction. But we'll save that story for some other time.

[7] Can you use the word "reams" when referring to the Internet? After all, it isn't really on paper, is it?

No More Secrets!

One place I found said that regenerative braking is the well-kept secret of motor control. However, when I learned the truth, I think it is just poorly explained. Let's start with the following, a diagram of a simple PWM controller for a DC PM motor.

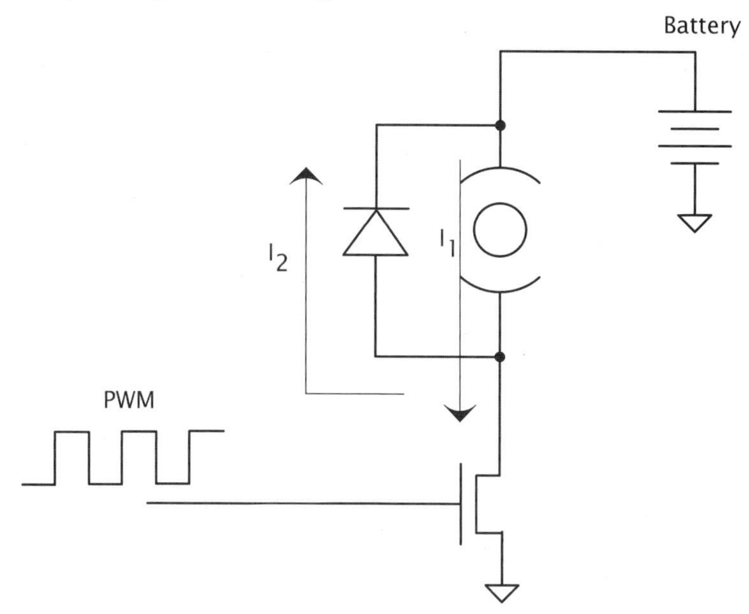

Figure 4-16 *PWM motor control*

A PWM is fed into a switch, such as a MOSFET, at a frequency that is high enough to keep current flowing in the inductor inside the motor, not at all unlike a switching power supply. When the PWM shuts off, the current flows through the diode, also known as a freewheeler. But the question that I kept asking myself was how do you get a motor, which is spinning at a lower voltage than the output of the battery, to push current back into the battery?

Let's start with a small change to the circuit above. We will replace the diode with a synchronous switch that goes off when the primary goes on, and vice versa. For the purpose of this discussion we will ignore the fact that the FETs need particular driving methodologies for the high side and the low side of a motor.

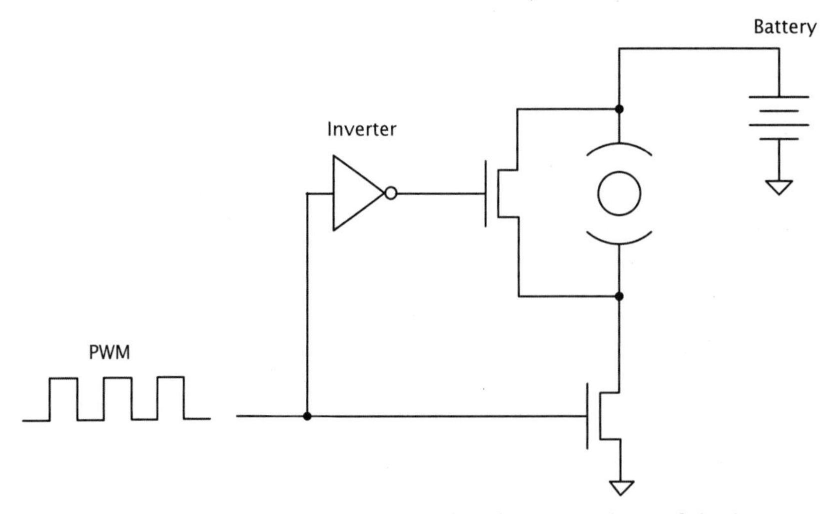

Figure 4-17 *PWM motor control with FET in place of diode*

I had read about this topology many times. It is usually brought up as a way to make your controller more efficient in terms of heat loss. This is because the FET has a significantly lower voltage drop across it than the diode does. I had no idea that it also functions as a regenerative brake, until now. Here is how it works.

A Little Elaboration

Keep in mind that in the original version of this controller there is a voltage across the motor that depends on the duty cycle of the PWM, but it is referenced to the positive output of the battery, not the negative side. That helped me to keep it in perspective.

Assume we have a 12V DC battery and there is 6V DC across the motor. That means you would see an average of 6V DC from the bottom of the motor to ground. Now let's say you spin the motor faster than 6V, say 7V for example. If you keep the same average voltage at the bottom of the motor, then you will have 1V extra to dump into the battery. This explanation doesn't entirely jive, but I think it will get you in the right frame of thinking. If you follow it to its conclusion, you will think that the version above with the diode should also regenerate, but it does not.

Let me elaborate. With the diode version, there is no braking force generated. That comes into play when the diode is replaced by the FET. When the freewheel FET turns on, the voltage generated by the motor is shorted back into itself. This provides the braking force and a current flow in the opposite direction through the motor. Remember the rule of inductors (since there is a decent-sized inductor in the motor). Once a current is flowing, it doesn't like to stop. So when the high side opens and the low side closes, current is now pushed into the battery and, voila, you have regeneration!

Regeneration Ain't So Bad

It turns out that regeneration isn't so tough at all. In fact, it is almost a side benefit of making your controller more efficient if you want to look at it that way. Now if there were just some way of making it more than 100% efficient, hmm...

Changing Directions

In a DC PM motor it is fairly easy to change directions of the armature. You simply need to reverse the voltage to the motor leads. A common way to do this is known as an H bridge, for the way it looks when drawn on a piece of paper.

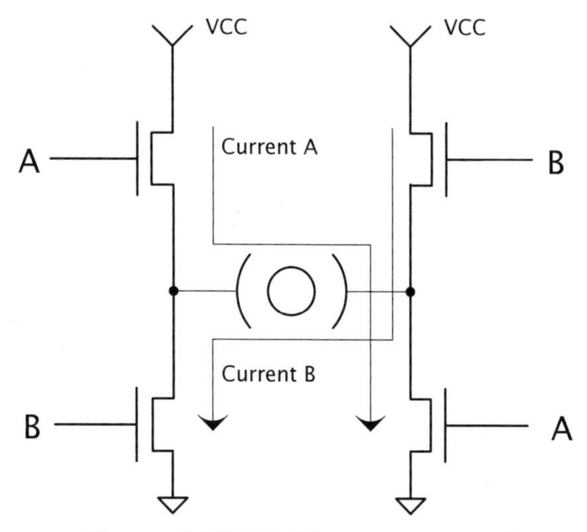

Figure 4-18 *H bridge motor control*

As previously discussed, the H bridge can be a linear or switch mode design. The same theory is applicable but it does become more complex. For example, you don't want to turn on both legs on one side of the bridge as you will create a short across your power supply. This is known as *shoot through* and will usually cause copious emissions of the magic smoke.

Synchronous switching of the opposing high and low legs with the appropriate devices (like FETs) will create the braking/regenerative effect that we have already mentioned.

Voltage and current feedback become more complex due to the fact that one leg of the motor is no longer tied to one spot all the time.

Conclusion

Controlling motors is one of the most complex and rewarding things you will do in electrical engineering. Setback and frustration will be rewarded with the pride of seeing something move! There is no way in this text I can possibly cover all aspects of motor control. I do hope, however, I have given you enough basic understanding that when you tackle motors and do more research on the topic you will be able to understand what you find out there.

Some Other Types of DC Motors

You will run into many types of motors. People have been goofing around with different ways to make a motor nearly as long as they have been messing with electricity. Here is a bit of overview on some various types of DC motors.

Brushless DC Motors

Brushless DC motors are a cousin to the DC PM motor we discussed above, but instead of using brushes for commutation, they usually use some type of electronic control. To accomplish this, usually the inside of the motor has the permanent magnets (where the armature is in the DC PM motor). This is known as the rotor. The windings are on the outside and are usually referred to as the stator, or field windings. There is no requirement for the magnets to be on the inside and the windings on the outside, but the windings are stationary and the magnets rotate. The rotor is turned by switching the stator windings on and off in a sequence that creates torque on the rotor. This is known as electronic commutation as opposed to brush commutation, as we have already learned about.

Figure 4-19 *Cool little brushless DC motor out of my RC airplane*

Often you will be told how a DC brushless motor is so much more efficient than a DC PM brush motor. There is some hype to dig through here. While this can be true, it is not entirely due to the fact that it is a brushless motor, as the brushless motor guys would have you believe. You will see numbers showing improved efficiency, but that is generally due to the choice of magnets. Most brushless motors are using rare earth magnets that have a much higher flux density than the more common ceramic kind. What this results in is less turns of wire for the same torque and speed. Less turns of wire means shorter wire, which means lower resistance. Since the resistance of the windings is the largest loss in the motor, this makes the motor more efficient.

DC PM motors commonly use the ceramic magnets, resulting in more turns of wire. To make them more efficient, you need to increase the wire diameter to lower its resistance. It is possible to use the stronger magnets in a DC brush motor. The most common place I have seen this is in the hobby store. There are some pretty cool motors like this for RC airplanes. When built with these "super" magnets, the DC PM motor is pretty close, if not better, to the same efficiency as the DC brushless motor.

Assuming good bearings, the next point of loss in a motor is in commutation. In the DC PM motor the brushes and brush contacts are the method of commutation. This interface is not perfect and creates a resistive loss. In the DC brushless motor, commutation is done with some type of silicon switch such as a FET, for example. Typically it takes at least six of these parts to commutate a DC brushless motor. These FETs have a resistive component (RDS on) that causes loss in the form of heat.

The biggest advantage to a brushless motor is right there in its name. It has no brushes. The brush in the DC PM motor will nearly always be the first thing to wear out. Brushes by their nature are designed to wear out, but don't let that stop you. There are many types of brush motors available and often they will be just fine for the application.

One thing about brushless motors to note is the controllers are more complex, requiring three to six times the power devices that brush motors use. But once you have them under control, you have already spent most of the money needed to make them go both directions. So if that is a feature needed, it could make a brushless motor more of a candidate.

Stepper Motors

Stepper motors are a type of DC motor in which the output moves a specific distance each time you energize a winding. They are a cousin to the brushless motor and a weird animal called the switched reluctance motor[8]. The ability to move a specific step makes them commonly used in positioning mechanisms. Printers use them by the bucket load.

Positioning is relatively easy since you can energize the windings and count the number of steps you have made to determine where the motor shaft is.

Stepper motors are characterized by their moving torque and holding torque. This is important to know because if you exceed either, your motor may slip and that would cause your count to be off.

---

[8] Somewhere between an AC motor and a brushless DC PM motor lies the switched reluctance design. It is rare enough that the reader is left to his own resources to find out how this puppy works.

Thumb Rules

- Linear controls cause less EMI.
- Linear controls are simple and cheap.
- Linear controls are less efficient due to heat loss.
- Switching controls are more efficient.
- Switching controls cause more EMI.
- Switching controls are generally more complex and expensive.
- Constant voltage makes for constant speed.
- Constant current makes for constant torque.
- Don't forget freewheel diode in a switcher.
- Replace the freewheel diode with a FET and you have a brake.
- Use an H bridge to change directions.
- Brushless motor controls are inherently bi-directional.
- Stepper motors move in small steps or increments.

AC and Universal Motors

Long ago a smart guy by the name of Tesla helped us all by convincing the powers that be that we should have an AC means of power distribution (vs. the DC local generators that Edison wanted). One key factor that helped with this debate was his invention of the AC motor.

There are many types of AC motors. One of the most common and the one we are going to review here is the AC induction motor.

An AC induction motor induces a current in the armature by varying the magnetic field in the stator. This induced current in turn creates a magnetic field that causes the rotor to turn, pushing against the first magnetic field. When I first learned this, it seemed to me that an AC motor can pick itself up by its bootstraps, so to speak. One result of this is that the motor tends to have a "sweet spot" where the rotational speed is just right, generating maximum speed and torque. At lower speeds the torque drops off pretty fast. This leads to the fact that AC motors are not known for low-speed torque (unlike the DC versions we just discussed). Due to this and the fact that AC motors run off a sinusoidal alternating signal, a huge percentage of AC motors are fixed-speed outputs where the speed depends on the frequency of the AC signal. There are variable frequency drives, similar in architecture to DC brushless drives, that can vary the frequency into an AC motor, creating a variable speed AC drive. Since AC motors do not have such a simple torque speed curve, these controls can be fairly complex, often using DSP chips to handle all the math needed to get what you want out of one of them.

AC motors have been around for years, making them relatively inexpensive, and their lack of brushes makes them last a long time. They can be built synchronous like a stepper motor so that you know they have moved a set distance every cycle of the AC wave. You will see them in all sorts of places, running compressors in a refrigerator to timing the icemaker circuit in the same fridge. Back before the "day of the diode" they were used in millions of clocks.

Universal motors are like a PM motor without the permanent magnet. They use windings with current flowing through them in the outer field instead of said magnets. What makes them universal is the ability to wire them to work with an AC or DC source. I shocked myself more than once rewiring the motor down in the old milking barn trying to figure this out.

Motors of all shapes, sizes, types and voltage preferences are out there. Hopefully I have provided enough background so you at least sound smart when asked about this topic.

Solenoids

The solenoid is an electromagnetic device that moves typically to only two positions. Akin to the stepper motor, they are rated by holding force and moving force.

Take a coil of wire and an iron rod that just fits inside the coil.

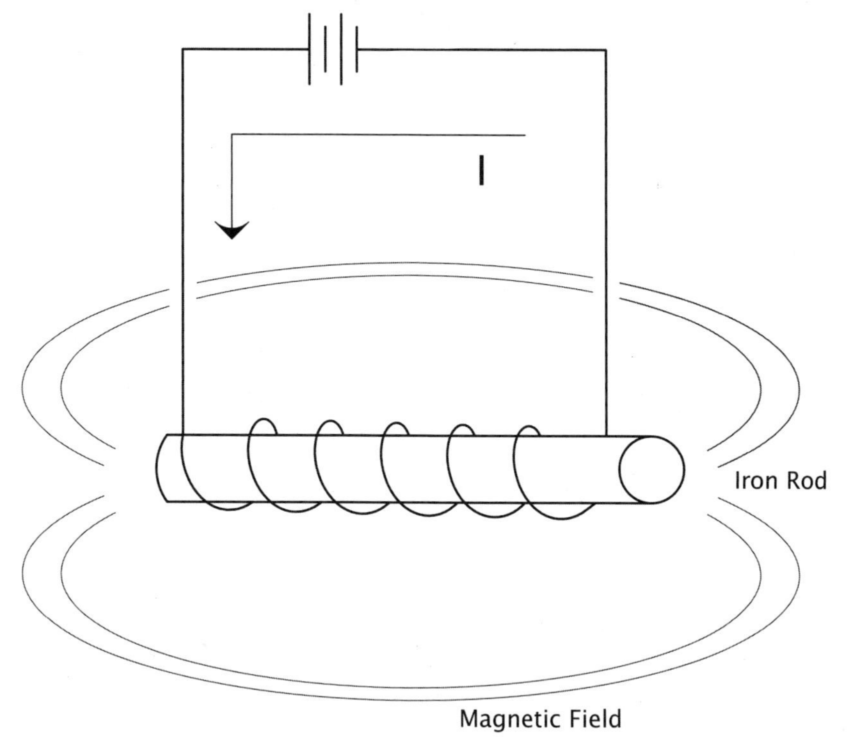

Figure 4-20 *Iron rod in coil of wire*

Energize the coil. The rod will center itself in the coil due to the magnetic flux running through it. It is in fact reluctant to leave the warm home of its cozy little coil. This tendency for ferrous material to align itself with magnetic fields is known as *reluctance.*

Shut the magnetic field down and the rod moves easily. Usually a spring is used to push the rod out of its cozy coil shell until power is switched back on and the rod returns to showing its reluctance yet again.

Solenoids are great if you need a short linear motion controlled by something electronic. Also, they are the basis for an electromagnetic cannon that we just don't have time to cover in this text. Too bad...cannons are fun.

Relays

Relays don't actually make anything move except the part inside them that closes a switch, so they may seem a little out of place in this discussion. However, they are definitely electromechanical in nature and I couldn't think of a better place to talk about them. Relays are very tough; they predate the transistor by a long time and are still in use. That should say something. They are basically the combination of a solenoid and a switch. A magnetic force pulls the switch shut or open depending on the particular device. Markings on a relay usually indicate the coil that operates the relay and the labels NO, NC and C. These abbreviations, also sometimes seen on switches, mean Normally Open, Normally Closed and Common, respectively. NO and NC refer to the state of the switch when the coil isn't energized. Common is a connection to both these switches.

There are two important specs on a relay—the coil voltage and the contact ratings. If you underdrive the coil, you might get the switch to close but there are no guarantees. Contacts are often rated at a minimum as well as a maximum current. Most engineers are diligent about paying attention to the max current, but often ignore the minimum current. Many relays used in a power setting (which is very common these days) rely on a certain amount of current to be present when the switch opens. This current creates an arc that cleans the contacts and keeps them from corroding. Do you have a relay that just stops working after a while? Chances are you are not meeting this spec. Use a relay in your design and you get to hear that satisfying click, letting you know that something is really working in that magic box.

Catching Flys

One thing all these motors, solenoids and relays have in common is the coil of wire that is switching current at some point. A coil of wire is an inductor and an inductor doesn't like current changes, does it? So what happens when you shut the current off in an inductor? As the magnetic field collapses when you cut off the current, a large voltage spike is generated. This spike is sometimes called the fly-back. To keep this spike from damaging components and to use the energy in it, most applications employ a fly-back or, as it is also called, a freewheel diode that shunts this spike back to its source.

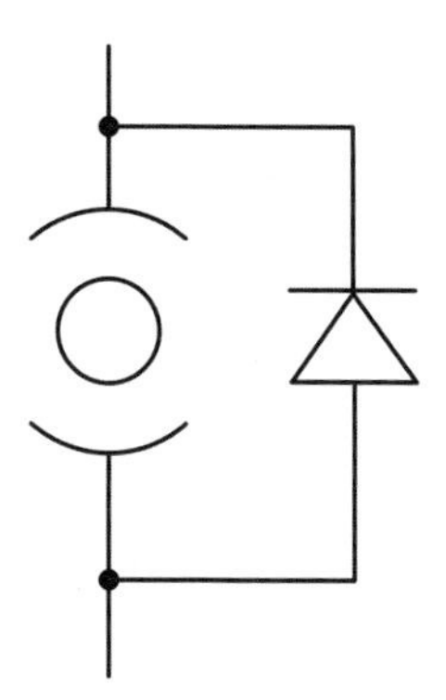

Figure 4-21 *An example of a fly-back diode around a motor*

If the response of the LR in this circuit is slower than the switching frequency, the diode acts as part of a filter keeping current moving through the motor. It smoothes out current changes which in turn smoothes the torque changes. (Remember how torque is proportional to current?)

In other cases, this diode may simply be capturing a transient signal to prevent circuit damage. Here is an example of using a diode to protect a relay circuit.

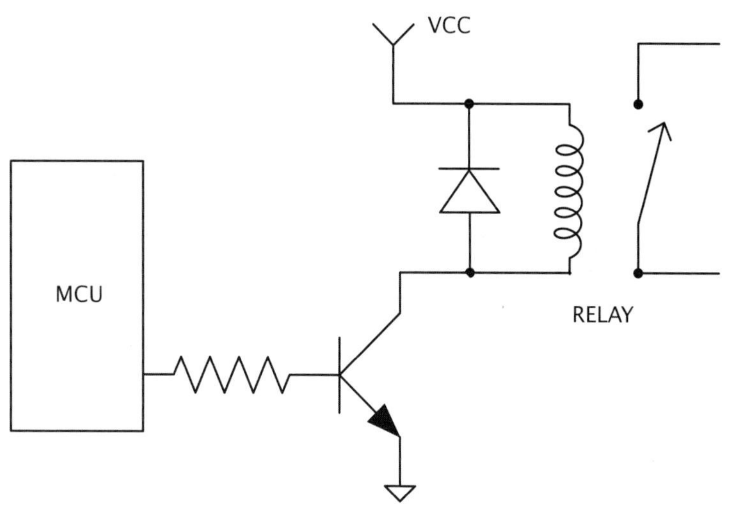

Figure 4-22 *Fly-back diode on a relay coil*

You can see that the voltage spike, inductive kick, or back EMF as it is called never gets over –0.7V because once it does it forward biases the diode and current flows back into the other end of the inductor. Now you know how to make a fly catcher out of a diode.

More Thumb Rules

- When the thumb rules go on and on, break them into smaller, more digestible pieces.
- AC induction motors induce a current in the core which in turn creates a magnetic field that turns the shaft.
- Universal motors can be wired for AC or DC.
- Solenoids are reluctant to leave the cozy coil cave when current is on.
- Pay attention to the minimum switching current on relays.
- Catch your flys with diodes to keep voltage spikes out of your circuits, unless you are trying to make a shock box to surprise your buddy.

Power Supplies

Whatever you do with electronics, you are going to need power to accomplish it. It will be useful to understand the basics of power supplies, as you are nearly guaranteed to deal with them at some point in your career.

It's All About The Voltage, Baby!

Most devices today want to keep the voltage constant. This means that current can vary as needed. In the world of power, particularly as it relates to the ubiquitous IC, it often seems that you never have the exact voltage you want.

A huge number of products run off of 120 VAC out of a wall socket. Another huge group runs off of batteries that are charged from those wall sockets, and another significant number runs off of batteries that you can buy by the caseload at any supermart. Just ask yourself, how many batteries did you buy last Christmas?

The problem is that most ICs these days want 5, 3.3, or even 1.5V DC. This is no where near 120V, and definitely not AC! Enter the power supply. They come in two flavors, linear and switcher.

Linear Power Supplies

AC rules! It is everywhere. It may seem like the world runs on batteries these days, but AC still has the majority foothold. Back when Edison and Tesla argued over what type of electrical power distribution we should have, I'll bet they had no idea of the type of integration that would occur in the world of electricity over the next 100 years.

One thing they did know about was the transformer. The basis of the transformer is AC current. Put AC into one side of the transformer and, depending on the ratio of windings, you get AC out the other side. So, put 120 VAC into a 10-to-1 ratio transformer and you will get 12 VAC out (minus heat losses due to the resistance of the windings).

The basic transformer is a very simple design. It is coils of wires on hunks of metal. That makes it robust. A transformer is a perfectly acceptable way to change the voltage of an AC signal. Transformers are used to jack the voltage way up to minimize losses over long wires, and then they are used again to bring the voltage back down to something safer to bring into your house.

They further knock the voltage down again in millions of products, but at that point they still output an AC signal. However, most of our chips want a DC signal, so what happens next?

It goes through a rectifier. There are two commonly used options, a bridge rectifier as shown in Figure 4-23, and a center tap rectifier, as shown in Figure 4-24. Note how this uses two fewer diodes and another wire to the transformer, yet the rectified output is the same.

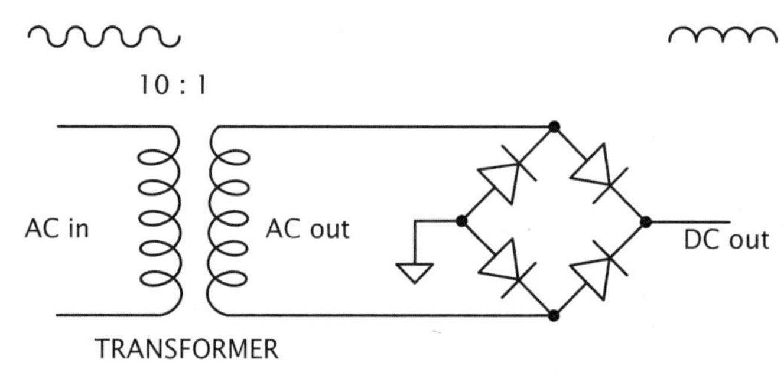

Figure 4-23 *Bridge rectifier*

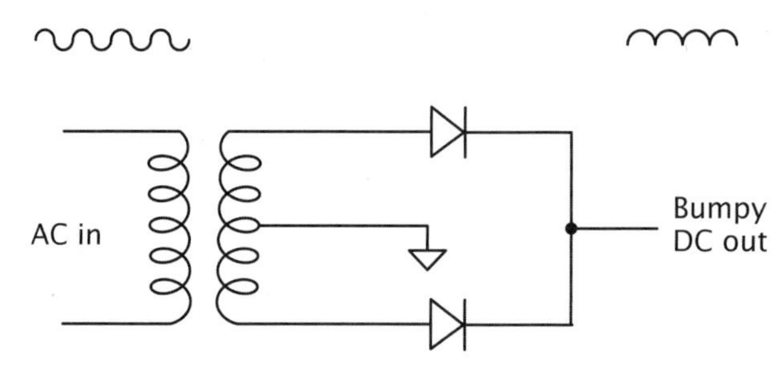

Figure 4-24 *Center tap bridge rectifier*

The output at this point is still too "bumpy" to be of much use to our sensitive DC circuits. The next step is to add a large filter capacitor to smooth out the bumps, as seen in Figure 4-25.

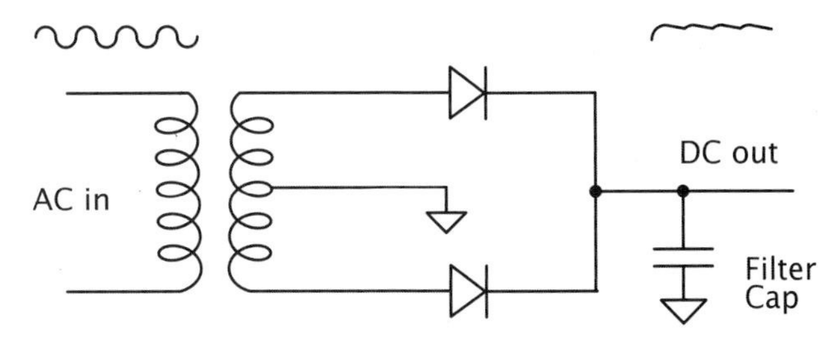

Figure 4-25 *Center tap bridge rectifier with cap filter*

Here you need to understand the principle of output impedance. Every power supply has it. The more current you pull out of the circuit, the bigger issue the output impedance is. Remember that Ohm's Law says that, as current increases through an impedance, the voltage drop across it increases. This means that the voltage at the output will drop as load increases. To further complicate things, due to the rectifier in this circuit you will see an increased ripple voltage on the output as load increases. So there are two important things affecting the voltage on the output of this circuit: the voltage going into it (which on most AC circuits can vary 10% or more) and the amount of current being drawn, increasing the voltage drop and the voltage ripple.

This is important to know as we feed this into the next part of the circuit, called a regulator. The regulator is a part that adjusts its output to maintain a constant voltage in the face of a changing load and a changing input voltage.

A linear regulator typically has a voltage reference (like a zener diode) inside it running on a small current that isn't disrupted by the load. It uses this reference and a negative feedback loop to control a transistor or other part inside to maintain a constant voltage at the output. This gets you to the nice DC voltage that your IC wants. The whole circuit from the wall looks something like this:

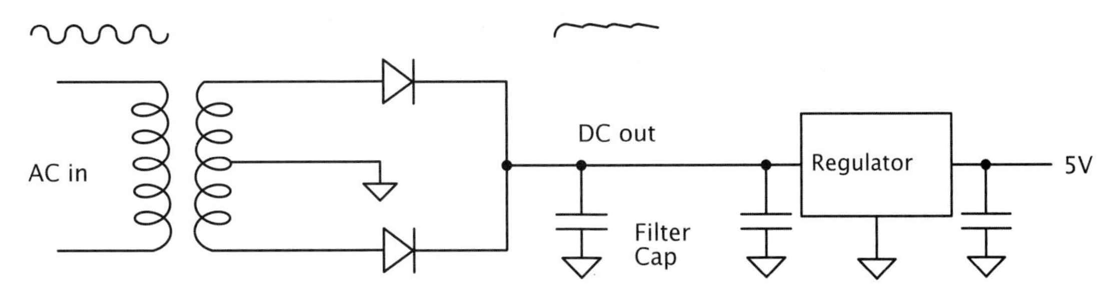

Figure 4-26 *Typical linear regulated power supply*

There are a couple of important things to know about linear regulators. They have a minimum input voltage. If the input voltage falls below this rating due to circumstances described above, the output will fall out of regulation. If this happens you can get ripple on the power supply to your chip. If it is small enough, you may not ever notice it, but if you have some high-gain circuits picking up AC noise, check out the power supply for problems first.

The other often-overlooked important spec is the power rating of the regulator. A regulator can only handle so much power, even with a heat sink. The power being dissipated by the regulator is the current times the voltage drop across the regulator, not the voltage at the output!

There are many other specs you should review in the datasheet, but these are the most important and often overlooked. Check them first. You can use linear regulators in any DC in, DC out situation. They will do very well in most cases and, to top that, they are very simple and robust circuits. Use them whenever you can. There is nothing wrong with this technology in certain applications, but if you need more efficiency, or maybe less heat, you should consider a switcher.

Switcher

A type of regulator and power supply rapidly gaining footholds over the older, linear designs is called a switcher. As implied by the name, it regulates power to a load by switching current (or voltage) on and off. For this writing we will focus on the current method. (Don't forget, however, that current and voltage are invariably linked as Ohm proved so well.) The secret to these supplies is the inductor and the secret to understanding an inductor for me is to think in terms of current. In the same way a capacitor wants to keep voltage across it constant, an inductor wants to keep the current flowing through it constant as well.

DC is What We Start With

Switching power supplies are DC-to-DC converters. Even those that have an AC input create a DC bus, using a rectifier circuit before implementing a switcher. You will see switchers replacing just the regulator in the circuit above working off of a DC bus voltage that has already been stepped down by a transformer. You will also see switchers that use rectified voltage right off the AC line and drop and regulate all in one step from 120V down to 5V.

The most basic current-switching supply I know of is the buck converter. A buck converter will knock a DC voltage from a higher level to a lower level. The following diagram shows the heart of a buck circuit.

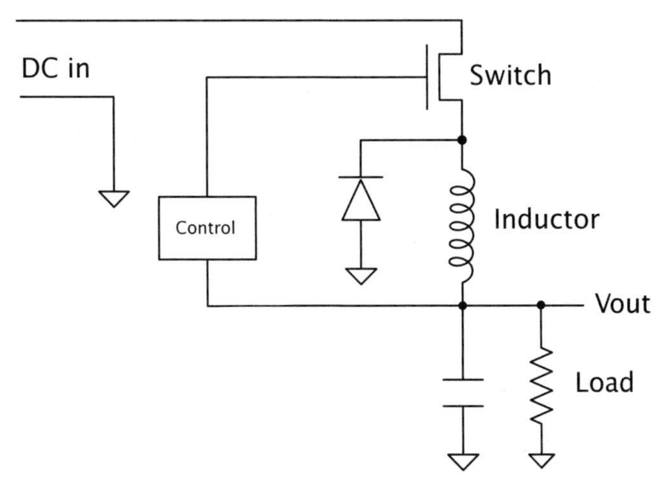

Figure 4-27 *Basic switching buck converter*

First identify four main parts: the inductor, the switch, the diode and the load.

Flip the Switch

Let'ss start with the load and work our way backward. To begin with, switching supplies like to have a load. Without a load, funny things can happen, but more on that later. What the load wants (in most cases) is a constant voltage. If I remember Ohm's Law correctly, one can control the voltage across a resistor (i.e., load) by controlling the current through it, so let's consider the flow of current in this circuit. We will begin with the switch closed. With the switch closed, current will flow through the inductor into the load. The current will rise based on the time constant of the inductor and the impedance of the load. Since the current rises, so does the voltage across the load. Assume now that we have a circuit that is monitoring the voltage across the load, and as soon as it gets too high it opens the switch.

Now what happens? First, remember this fact. Just like a capacitor resists a change in voltage, an inductor resists a change in current. When the switch opens, the inductor tries to keep the current flowing. If there is nowhere for it to go, you will see a large voltage develop across the inductor as the magnetic field collapses. In fact, at time = 0 the value of this voltage is infinite or undefined, whichever suits you. That doesn't happen in this case due to the diode and the load.

The current flows into the load, and the reason it does so is because of the diode. Consider it this way—current wants to keep flowing out of the inductor and into the other side of the inductor. Without the diode there would not be a path for this current to follow. However, with the diode, this current is pushed through the load. So now the switch is open, and current is still flowing into the load. This current starts out at the same level it was when the switch opened (an inductor wants to keep current the same, remember!) and it decays from there. As the current falls, so does the voltage. Of course we still have a circuit monitoring the voltage across the load, and as soon as it gets too low, it closes the switch again.

Voila, the process starts all over. There are two important things I noticed once the pieces fell into place in my head. The first is this control circuit I just described can be implemented with a simple comparator and a little hysteresis. Of course, that would lead to the frequency of the switcher being determined by the value of the inductor and the impedance of the load. That may or may not be a desirable trait. The other thing I realized was that when you first turned it on, the circuit would want to slam the switch shut and keep it there for a long time while current builds up in the circuit. Are you beginning to see why switchers need a load?

Luckily, others much smarter than I have dealt with these problems already. That is why you hear terms like "soft start" and "built in PWM" when you start studying switching supplies.

Some Final Thoughts

Since designing switching supplies, getting them stable and dealing with the inductor specs can be a bit demanding, technical and tedious, all sorts of industry help has sprung up in the effort of various companies trying to get you to use their parts. You will find design guides and even web design platforms out there to help you build a switcher for your design, and I highly suggest you take advantage of them.

These days you will often find all the brains, switching components and feedback parts in one cute little package[9], making the design nicely compact and small. You can make switchers that boost voltage as well as the buck versions, and some that even go both ways, but ultimately they rely on the fact that the inductor wants to keep current flow the same. We will save the more in-depth review for another book on another day.

The best thing about switching supplies is the fact you can get by with relatively little copper and attain very high efficiency (meaning less heat). The reason for this is that the decay rate of the current in the inductor depends on the size of it, but if you switch it faster, the average current and thus the average voltage is still maintained. So you can get by with much less copper, especially for larger current draws at low voltages. The efficiency is good because much less power is spent heating the copper in the small inductor than in an equivalent transformer design. However, all this comes at a price. Switchers are known for their high-frequency noise that has disrupted many a sensitive analog design. But who cares about analog any more, right?

Thumb Rules

- Make sure the lowest dip on the ripple voltage doesn't go below minimum input of the regulator.
- Check your supply at ±15% of the AC input signal.
- Linear regulators dissipate heat/power based on the current times the voltage from input to output (i.e., across it).
- Switchers exploit the fact that inductors want to keep current flowing even when the switch is open.
- Switchers are more efficient and create less heat but generally are more finicky to set up.
- Linear supplies are very quiet when it comes to EMI.
- Switchers tend to be very noisy when it comes to EMI.
- Switchers need a minimum load to work correctly.

[9] I know, only real nerdy engineers would think an IC could have cute packaging, but I have never denied my nerd-hood.

When Parts Aren't Perfect

Before we get into the problems that parts can have, we need to introduce the concept of an equivalent circuit. It is pretty simple: to create an equivalent circuit, you represent all its idiosyncrasies with combinations of perfect components. This is good for two reasons. First and most obvious, it makes it possible to model the effects of the imperfections. Second, and most important in the World of Darren, is that seeing the combinations of the parts that make up a real component makes it easier for you to apply the basic understanding of the perfect parts to grasp what the real part is doing.

Everything Is Everywhere

The basic three electrical components are like sand at the beach. They get into everything. In a way they are more prolific than sand in your sandals since the effect of one basic component can be found in another. This fact is one of the most common causes of error you will have between how the equation predicts a circuit will work and how it actually operates. Chalk this up as one of the reasons datasheets are so important, even the ones that describe the most basic components. Datasheets will characterize the components, describing these error sources, such as the equivalent capacitance that may be found in a reverse-biased diode.

Most texts call these effects "error sources" since they are what makes the difference between a perfect or ideal component and what you actually have to work with. There are other types of error sources in every component and we will discuss a few of them later on, but those pesky R, L and C in some combination or another are pretty much everywhere. (I hope the reason for drilling the basics of these components is becoming more and more clear. It is appropriate to experience an Aha! moment right now and say to yourself, "Now I see why I need to know those basic parts by heart!")

The most general guideline to follow when you are looking for error sources is to ask yourself the following: Is this error source enough to account for the effect I am seeing? Let's go back to the diode example for a moment. A diode has a bit of capacitance when it is reversed biased, typically in the picofarad range. Consider this circuit.

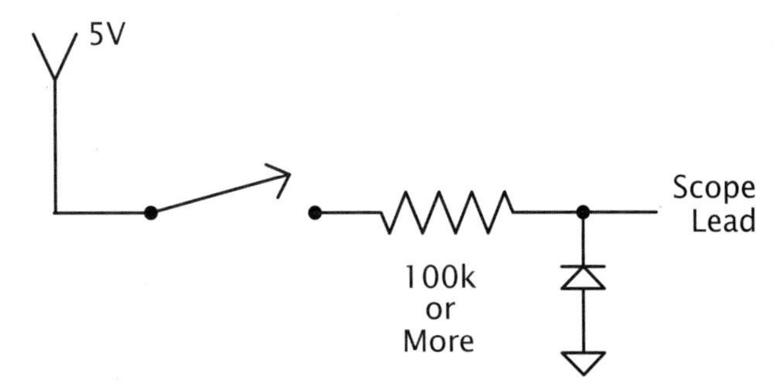

Figure 4-28 *Circuit to examine*

If you hook your scope lead to the output, and flip the switch, you see the following:

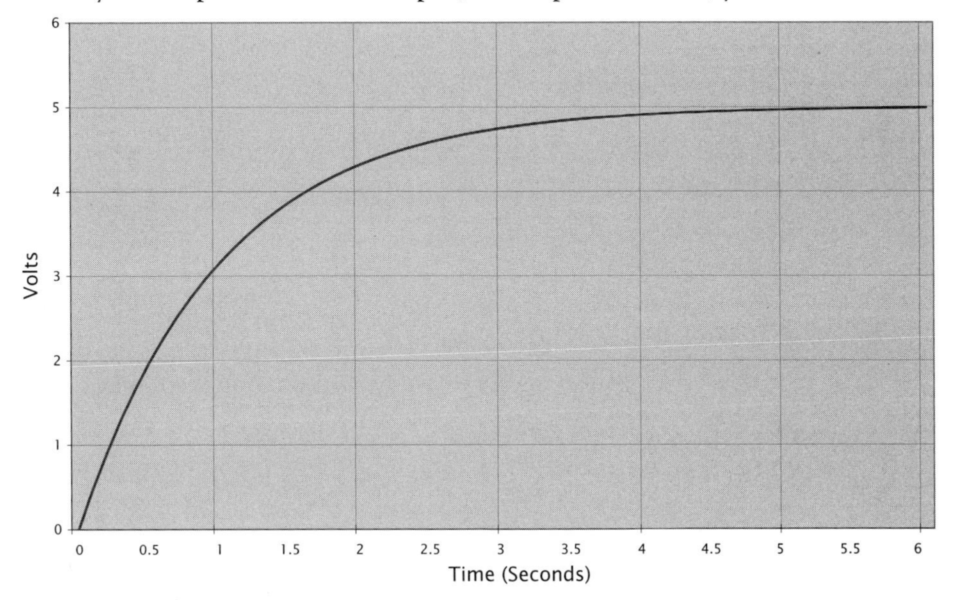

Figure 4-29 *RC curve seen on your scope*

Since there is capacitance in this diode, an RC curve is what you would expect to see in a situation like this, but is it really due to the error source in this diode or is it caused by something else?

Here is where the datasheet comes in; looking at the specs on the diode you are using, you find out that this capacitance is typically 100 pf. Plug that into the equation for the response time of an RC circuit,

$$\tau = RC \qquad \text{Eq. 4-4}$$

The number that pops out is **10 μS.** Taking a look at the scope again, you now pay attention to the time it takes for this curve to get to about 63%, remembering that is about how far this curve gets in one time constant, or tau. Being such an astute scope operator, you use the cursors on your tool and you measure a time of about one second for the signal to get a little over halfway to its final value. "That doesn't make sense," you should be saying to yourself, "If the diode is responsible, it would have to be much faster."

The moral of the story is to expect every component to have some amount of the basic three, but understand the magnitude so you can decide if it is causing the effect you are seeing.

Error Sources, Ideal vs. Real

In any circuit you design, there will be sources of error—things that just aren't perfect, sensors that are off a little, parts that aren't quite to spec, or any myriad of problems.

What do you do about it? Nothing, if the error source isn't causing you any problems. For example, a leaky cap may not really matter if you have plenty of power available. However, if the circuit is running on batteries you could have a problem on your hands. First and most important, determine if the source of error is an issue or not before you go about trying to solve it.

Once you figure you have a problem on your hands with a particular source of error, there are three ways to deal with it.

1. Get a better component.

   It's tough to plug the hole in a leaky cap[10]; it isn't like the boy at the dike—you can't put your finger in the hole and stop the leak. Sometimes your only choice is another component. In this case you might specify a tantalum cap instead of an electrolytic. Consider, however, that often the better compo-

[10] A very common source of error in a capacitor is a DC current flow. Remember the ideal cap will block all DC signals. You can think of it as a large resistor in parallel with a perfect cap. It is common enough to have acquired its own slang term. If this current flow is significant, the cap is said to be "leaky," as DC current seems to leak through it.

nent costs more, so spend wisely, not indiscriminately. Do note, however, that this is usually the quickest way to solve the problem as the design doesn't have to change.

2. Shore up the weak component with another component.

   For example, the frequency response problems with electrolytic caps can be dealt with by adding another cap in parallel, a smaller one that has no problems with higher frequencies. (You may have noticed regulator reference designs do just that to assure a stable output. Now you know why.)

3. Design the error out.

   This approach will take the most engineering effort, as the goal is to change the design so that the error is no longer significant. The proverbial op-amp is an example of this type of effort.

Now that we know how to deal with the problem, let's look at some common parts and typical sources of error. This will be an overview based on personal experience. It is no substitute for curing insomnia with a good datasheet.

Resistors

I would have to say that resistors are the most stable and predictable of the three basic components. Carbon film resistors have very little inductance or capacitance. It is rare you will have a problem with this unless you are dealing with radio frequencies and high clock speeds. In most cases the effect of the PCB design will be worse than the resistor itself.

The biggest issue with these common resistors will likely be heat. Exceeding or coming close to the wattage rating of these parts will make them vary significantly from their nominal value, so it is a good idea to give yourself plenty of headroom with these resistors.

Another common resistor typically used in higher power applications is a wire-wound coil with a ceramic block molded around it. In this case inductance can be a significant effect since there is a coil of wire and, as we know, coils of wire make inductors.

There is a whole industry of low-inductance power resistors that you can get to work around this problem.

Capacitors

In my limited experience, I have never seen a cap that even comes close to being perfect. A perfect cap would not heat up, but in fact they do. The natural conclusion you should come to is that capacitors have some type of resistive component. In fact they do and it is called ESR, or equivalent series resistance.

According to the equations, a 10-μf cap should have nearly the same impedance at 100K Hz as a 0.1 μf cap does, but alas this is not the case. That is why you often see a large cap with a small cap next to it on a power-supply circuit. Nearly all caps vary capacitance over a frequency range.

Big electrolytic caps often "leak" like a sieve. There is no particularly easy way to deal with that. You have to live with it or get a better part. Believe me, if you are trying to make a really low power circuit, the last thing you want is a cap spilling electrons all over the place.

One other thing I had to learn the hard way is that a cap only meets the rated capacitance when at the rated voltage. Sometimes overrating the voltage on the cap too much can leave you with a different capacitance than you expect.

Polarized capacitors will act like a diode if you don't bias them according to their markings.

Many caps will vary 20% over their temperature range; you may not want them next to a power resistor on your PCB.

You need to peruse the datasheets of caps carefully when you are picking them for a particular application.

Inductors

As these are most often coils of wire, you might imagine that resistance is one of the most common sources of error in an inductor, and you'd be right. Resistance is a major source of error in inductors. This usually causes heat and power usage that you may or may not want. Minimizing the current flow through the inductor makes the resistance less of an effect and is something you may be able to do at the design stage.

Many inductors are warped around some type of ferrous core. An effect called core saturation occurs when the magnetic field exceeds the amount the core can handle.

There are some capacitive effects between the coils of wire, but they are so small that we will ignore them in this text. If you are cranking out the gigahertz needed to make this important, you are probably reading a book about this stuff written by someone much smarter than I am.

Semiconductors

One of the things that every diode, and every semiconductor based on the diode, has in it is a voltage drop. For example, if your transistor amplifier doesn't see a base voltage over 0.7V, you won't get it to work.

Rail-to-rail op-amps are more expensive than their predecessors because they employ circuitry that eliminates these voltage drops so that outputs and inputs can get to their power rails.

In the datasheet of these parts, you should look for output impedances and capacitive effects. Inductive effects are generally small and insignificant in semiconductors.

Heat can also cause error in semiconductors. It generally affects the internal resistance and can cause avalanche[11] failures. It also seems to me that the most often overlooked part of the design is heat dissipation. The same engineers that can easily calculate the wattage needed for that specific resistor value will overlook the amount of heat dissipation in a semiconductor. Take the current though the part times the voltage drop across it and you will see how much power is being dissipated.

The world of semiconductors is so widely varied that there is no way this overview can be anywhere near comprehensive. I have to sound like a broken record (or should I say scratched CD?) and tell you to refer to the datasheet.

[11] Like an avalanche, when it starts to fail all heck breaks loose, usually resulting in an interesting smell.

Voltage Sources

What would cause a voltage source not to maintain the voltage output? Let me give you a hint, when put under load, a voltage source will heat up. So what creates heat? You got it, resistance. A voltage source has an internal resistance. A battery is a good example.

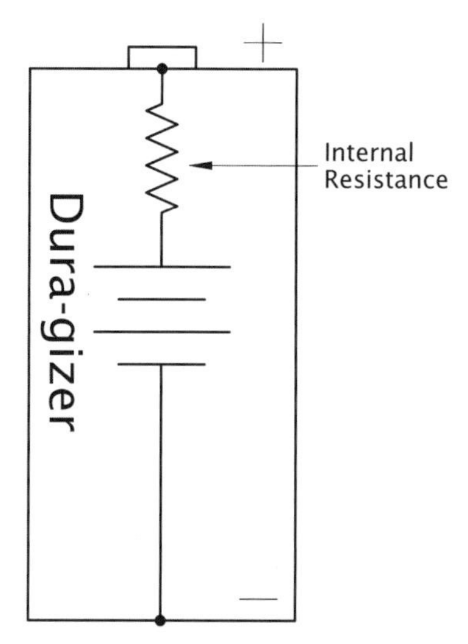

Figure 4-30 *The Dura-gizer, now that is one tough battery!*

As current is applied to the load, the voltage drop occurs across this internal resistance, just like the voltage divider rule says it will. This resistor inside heats up just like one on the outside does, making the voltage source warm. If the source doesn't compensate for it, you will see less voltage at the output.

Summary

Parts just aren't perfect, I have seen motor bearings wear out prematurely due to capacitive effects and seen caps overheat and pop their top. Truly the best thing to do is keep looking at the datasheet. Parts engineers do their best to characterize the deficiencies of the part and put it in the datasheet for you.

Thumb Rules

- Always ask, is the error source in this component enough to cause the effect I am seeing?
- If the source of error isn't large enough to be an issue, forget about it and move on.
- When fixing errors, get a better part, shore it up, or design it out.
- Caps vary with frequency.
- Inductors have internal resistance.
- Semiconductors have voltage drops and heat issues.
- Voltage sources have internal impedance.
- You can't study the datasheet too much.

Robust Design

Most engineers want to over-design, give themselves plenty of headroom, and use parts that are double or triple the spec they need. Usually the manager is there saying, "It needs to cost less or it won't sell" or "Do you really, really need that part?" To be honest, the answer lies somewhere between these extremes.

Can You Tolerate It?

Let's start with a completely general off-the-wall statement that you might hear from someone with a sharpened, somewhat devilish hairdo. "A robust design handles two things well, the inside world and the outside world." A look of consternation scrambles across your forehead. "What in the world does that mean?" you ask yourself. Let me proceed to enlighten you on this bit of pointy speak.

The inside world is all the parts that make up the design. In any production process, these parts will vary in specification. The question to ask is will the circuit operate correctly over the tolerance ranges of the parts? If the answer is yes, then the design is robust internally. The inside world is good to go. Don't assume, however, that only electronic components have tolerances. This point is best taught by example. In a design I worked on some time ago we were using an optic sensor to detect the motion of a belt. We correctly analyzed the tolerance of the opto, but as we began testing on a pilot run we discovered that the belt we were using varied in opacity. If the optic sensor was at the high end of its tolerance and the belt was at its most transparent, the signal from the sensor wouldn't get high enough to guarantee the logic input would read it correctly. In a production run, a problem like this would appear as a random failure. This type of failure is known as a tolerance stack up. It occurs when the stack up (the additive effect of the variations) of two or more components combine to create a failure. It is more difficult to analyze than a single component tolerance issue. Probably the best way to preempt this type of failure is with the help of simulators. Take caution though: make sure your simulation accurately represents the design with nominal perfect components before you start running tolerance analysis on it. (See the section on simulators for more suggestions.) The great thing about a simulator, though, is the ability to vary all the components over their tolerances and see the effects without building a whole bunch of parts. You can then adjust your design and component specs to increase the robustness of the product as far as the inside world is concerned.

However, the outside world is a different animal. A good design can handle the things the outside world throws at it. In the electronic realm all sorts of interference can disrupt your design. I once read an article that described something called a rusty file test. After the engineer was done with the part, he would plug it into the wall and plug in a home-built test fixture next to it that he had made. It consisted of a wire from neutral connected to a file. The hot wire had a bare end that he would proceed to rub up and down the rusty old file, sparks flying everywhere.[12] If the circuit passed this test without a hitch, he figured it was good to go. This is known as EMI or electromagnetic interference. It really is a whole topic unto itself, so I have dedicated a chapter to it. Skip ahead to it if you can't handle the suspense!

Don't limit your focus on the outside world to only electrical interference. There are many cases where other things can cause a problem. Vibration, for example, can cause traces on a PCB to crack and solder joints to become faulty. Increased humidity can swell a cheap PCB, causing mechanical deformation and cracked connections. It can also combine with debris to create electrical shorts on circuits that you don't want shorted. Temperature can be particularly tough on electrical components. You should review the temperature range your circuit will be subject to and compare that to the specs in the datasheet. Don't forget to include the operating temps of the device you are using in this analysis. For example, power components usually get pretty warm just operating. Toss them into a 70-degree ambient and you could easily push them over the max temperature spec.

How do you go about making your design robust externally? There are several approaches to take.

The most important, in my opinion, is doing everything you can in the fundamental design to get it to handle the environment it is in. Often a few changes to the PCB layout itself can make a circuit handle EMI better than putting all the shield-

[12] It was written by Ron Mancini in EDN, but I have to say do not try this test at home. There are much safer ways than the procedure described; I mention this because it creates a vivid picture of the junk out there that is trying to mess up your circuit.

ing around you can fit. Larger traces can combat mechanical deformation, and a few well-placed holes can help manage temperatures.

Reading, reading and re-reading the datasheet for the component you are using is probably the next most important thing you can do. The more you know about the parts you are using, the better you will recognize things that might upset your design.

The third and most extensive effort that will help is to test, check, test, and re-test the design. You need to recreate the environments that it will be subject to and see what happens.

Now, to top it off, you can have a situation where the problem is a combination of the tolerance of the internal design and the environmental effects it is subject to. These situations are nearly impossible to predict and are often simply discovered in the course of business. There is only one thing you can do about that, figure out what is needed to prevent it, make the change and document it for future use on similar designs. I recommend that every engineer and engineering group keep a document of design guidelines where you write down those rules of thumb that you discover along the way. Don't just write it down, but read it regularly to keep those things you have learned fresh, as you do each new design. This alone can be a powerful tool. Some years back I took over an engineering group. When I first started managing it, it seemed like we were always being called to the production line for some weird problem or another. We spent more time chasing problems than engineering new products.

We began a focus on robust design principles, and one of the first things I implemented was the design guideline documents. Every time we found a new design rule to follow we wrote it down and referred to it regularly so it would be implemented with each new design.

Over about a three-year period, those urgent calls to production began to drop off. We went from spending over 50% of our time in production support to spending less than 10%, a couple of years after that, we were spending less than 1% of our time dealing with production problems. Considering we were moving tens of thousands of products out per day, that was a great achievement. Months would go by without a call, where before it was every day. When problems did occur you could nearly always trace back to a guideline that we had written down and simply neglected to follow. The hard part became referring back to those documents each time you created a new design. That being the case I suggest you try not to let them get too large. The bigger these documents are the less likely you are to read though them. So try to keep them to a few pages as they will have a tendency to grow a lot.

In an effort to quantify what the outside world can do, many standards have been written. They are some great yawners (meaning they will knock you out in about 5 minutes of reading) but they can give you some real insight into what your design will be subject to from the outside in. Documents like IEEE 62.41 that describes the world of EMI or UL 991 that describes how to make a control safe. The list goes on and on. Do a little research on what you are working on and see if someone has written something about it. If your boss doesn't understand the need for time to do this, show him this paragraph.

> *Boss, it may seem like nothing is getting done when the engineer is sitting there reading, but trust me, this effort can save you millions in production downtime, so give your engineer a chance to succeed and you will not regret it. Engineer, this doesn't mean you should just read and never design anything; I would limit this research to about a 10% to 20% ratio of design vs. research maybe more if you are talking something you have never done before.*

Reading these documents work particularly well if you are tossing and turning all night as you try to figure out what is wrong with your design. I would keep them by the side of my bed. That way I could learn some more for a few minutes and also get some sleep. They not only help with the design, they are a great cure for insomnia!

Learn to Adapt

Have you ever finished a product design after which some change was required that you desperately wished you had been told about at the beginning? Have you ever said, "If you had just told me sooner, this feature would have cost half as much to add on now"? You may even have had your boss say "Why didn't you do what I told you to do" (when you did exactly what he/she wanted). I'll let you in on a secret. Most of your pointy haired bosses don't want you to fail. They just want to ship a killer product so their bonus will be bigger.

They don't tell you sooner because they don't know sooner. They try to guess what the customers want and give it to them. In his mind he didn't say do such and such a product. He said make this product successful. As companies chase the market around, new products are developed, changed, and changed again. I call this Management Always chasing the Market Around or MAMA for short. (There is nothing like an acronym if you want a point to be remembered. I predict that some day in the far future acronyms will be the prime method of communication due to their efficiency!) Now because of MAMA, many engineers experience consternation when their product definition changes.

In the world of consumer products, this is bound to happen often. When I first took over the engineering group where I work, this particular frustration was often felt. As I worked with the various designers in the company, I found that it was possible to anticipate these changes and prepare for them. When you get good at this, you can respond to changes easily and quickly, and you can also develop a number of derivative products quickly and inexpensively.

Modular Design

One of the most important things that you need to do to anticipate change is to modularize your designs. Here, hardware designers can take note from their counterparts in software design. Good software engineers build blocks of code that can be used and re-used again and again. However, I often see hardware designers start with a clean sheet for every new design.

To make your modular design work for you, you must evaluate the products you are designing. Are there any components that are commonly removed and installed on various designs? What parts are common to all or most of your products? Sit down and draw lots of block diagrams and ask yourself, "Is this a part that needs to be easy to take on and off?" If it is, it may be a candidate for a separate PCB or a section of the PCB all to itself. In a line of stereo products, for example, you keep the tuner section separate from the pre-amp and so on. (On a side note, this often improves the robustness of a design as well.)

A great advantage of this modular approach is the way it can accelerate the development process, by using separate engineers on the various modules. It also allows you to upgrade or improve parts of the design without redoing the whole thing. But in my mind, best of all, it makes it easy to change a feature when your boss decides he really didn't want that there on this particular model.

One word of warning, however—you need to be careful what parts you choose to modularize. Too many modules can add up to extra cost in every product you ship. Make sure it makes sense.

Anticipate Changes

Try to get involved in the creation process so that you can see various phases of evolution the product design has gone through. Often, changes that are made will be back or forth on this evolutionary path. Keep asking yourself, where else could this be used? How would I change it to work there?

Look for places where a part seems to be missing. For example, you are asked to make a PCB with a row of LEDs that look like this:

Figure 4-31 *Row of LEDs management wants*

Say "Great, no problem," and then create a PCB with this row of LEDs and simply do not stuff the missing one for now.

Figure 4-32 *Row of LEDs you actually put in*

Do not hesitate to tell your co-workers or boss what you are doing. They can be a great asset in anticipating changes they might come up with later. The bottom line is that it takes a tremendous amount of work to redesign every product every time, but if you can develop effective modules and anticipate changes when you are engineering the product, you can bring things to market faster and cheaper than anyone else. A nice benefit of this type of anticipative design is that when you are asked to develop a similar product, you have all the pieces in place. You simply add or subtract the required feature and are done with it. And best of all, MAMA will not drive you berserk!

A Final Word of Caution

It is possible to go too far with this philosophy. Don't try to make your design so universal that it comes at the expense of getting the product to market, or adds so much cost for all the options that it is no longer viable. Remember, there is also a chance you will never use the option you built in, so choose wisely, young Jedi.

Thumb Rules

- Read the datasheet.
- Consider tolerances.
- Know the environment.
- Test, check and retest.
- Make your own list of thumb rules or design guidelines.
- Do research on standards or guidelines that exist for your product.
- MAMA can be frustrating.
- Modularize the design.
- Anticipate changes.
- Don't go too far.

Some of My Favorite Circuits

Every engineer has their favorite batch of circuits and I'm no exception. There are tons of circuit cookbooks out there that show how to implement no end of cool features. There are so many that you could spend all your time searching them and never getting anything done. I suggest you develop your own favorite basic circuits that you know well and intuitively understand. This is simply an extension of the "Lego" philosophy that we discussed way back at the beginning of the book. Here are few of my favorites. These are in addition to all the circuits I have used as examples up to this point, one reason they make such good examples is that they are so useful.

Hybrid Darlington Pair

Cool application note, using two transistors to switch a signal level Vcc PNP switched by NPN.

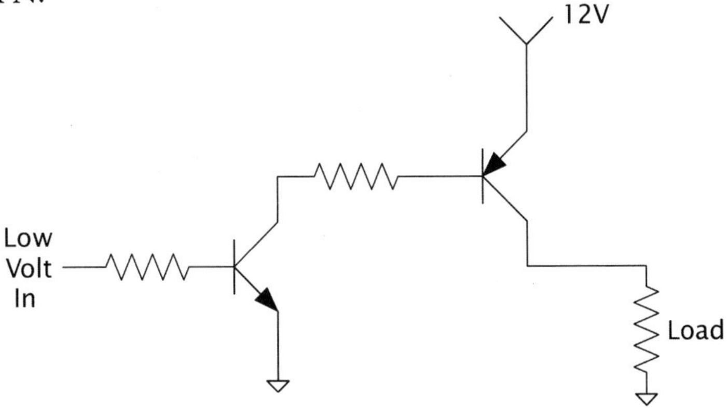

Figure 4-33 *Vcc PNP switched by NPN*

This is a handy circuit that switches a higher level voltage with a lower level one. Say, for example, you have a micro with a 5V output and you need to drive a 12V load. For a reason you can't change, you have to switch the Vcc leg. In this circuit you turn on one transistor with a 5V signal, which in turn activates the other transistor switching the higher voltage to the load. This works because the transistors are current driven; when you shut off the current flow to the PNP transistor, it shuts off regardless of the voltage. Another plus is that this circuit has Darlington-like properties without one of the downsides. You won't need a lot of current to the input to switch the output and unlike a traditional Darlington pair, the voltage drop across the output is much smaller. You don't have two series base junctions to contend with at the output. If you still don't follow, try a little ISA[13] on it.

DC Level Shifter

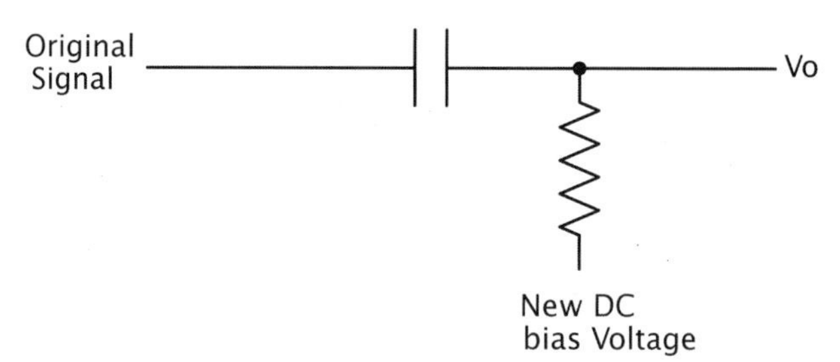

Figure 4-34 *Change the DC bias on an AC signal*

This is really the high-pass filter that we have already studied but with a slight twist. Instead of ground, we hook the resistor to a reference voltage. Since DC has a frequency of zero, only the AC component will pass and in the process a DC bias will be applied to the signal. Make sure you don't size the cap and resistor so that the signal you want is attenuated.

Virtual Ground

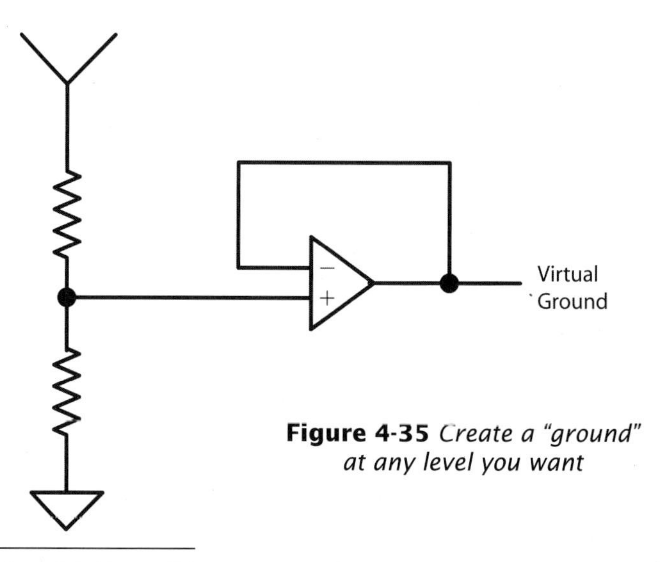

Figure 4-35 *Create a "ground" at any level you want*

[13] Intuitive signal analysis (ISA). I still hope to someday cement my legacy in an acronym.

Using the voltage divider as a reference, the op-amp becomes a voltage source with the output matching the voltage at the divider. This can be very useful when you are trying to handle AC signals with only a single-ended supply circuit.

Voltage Follower

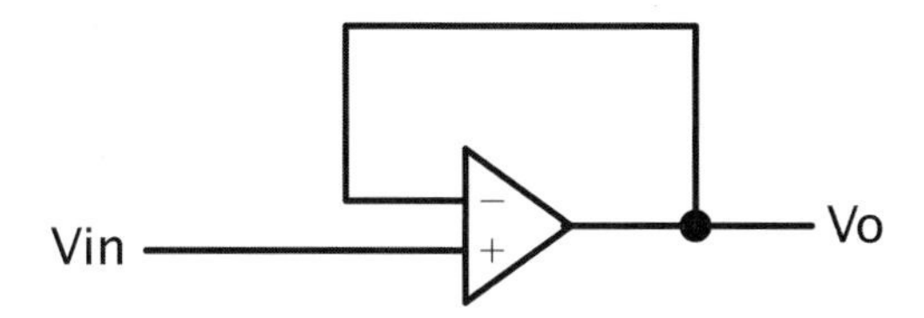

Figure 4-36 *Voltage follower*

This one is mighty useful when trying to measure a signal that is easily affected by load. $V_i = V_o$, but, best of all, V_i isn't loaded at all, thanks to the buffering effect of the op-amp.

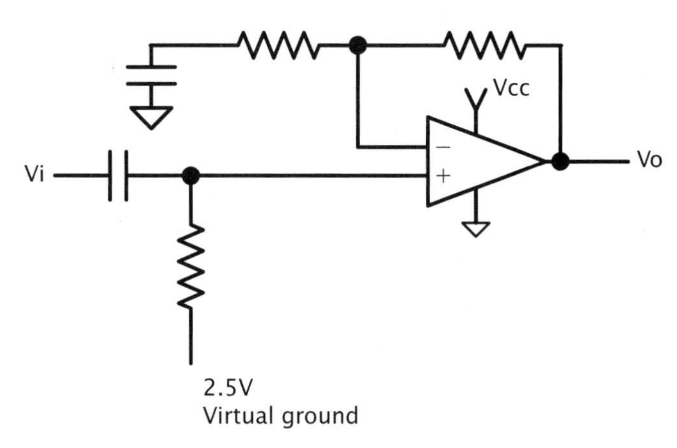

Figure 4-37 *AC only amplifier*

This is another great circuit that works nicely amplifying AC signals with a single-ended supply. It also has the benefit of not amplifying any DC signal components, keeping things like DC offsets from making your signal rail. This happens because of the cap in the feedback circuit. Since the cap only passes AC current, DC signals see that point as disconnected. When the resistor to ground is disconnected, the op-amp acts like the voltage follower in the previous circuit.

Inverter Oscillator

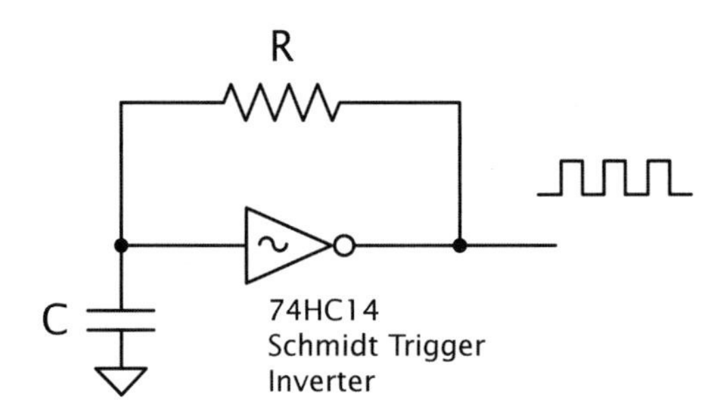

Figure 4-38 *Schmidt trigger oscillator*

I saw this in the back of a data book years ago; I think it was a Motorola logic data book. This was way back before the internet you—used to have to turn pages to find this stuff! The way it works is based on the fact that the Schmidt trigger inverter has hysteresis built into the input. This makes the output stick at a high or low level till the cap on the input charges to the threshold voltage that trips the inverter. Output flips and everything goes the other direction, repeating indefinitely. Adding some diodes to the charge and discharge path can affect the duty cycle of the output.

Constant Current Source

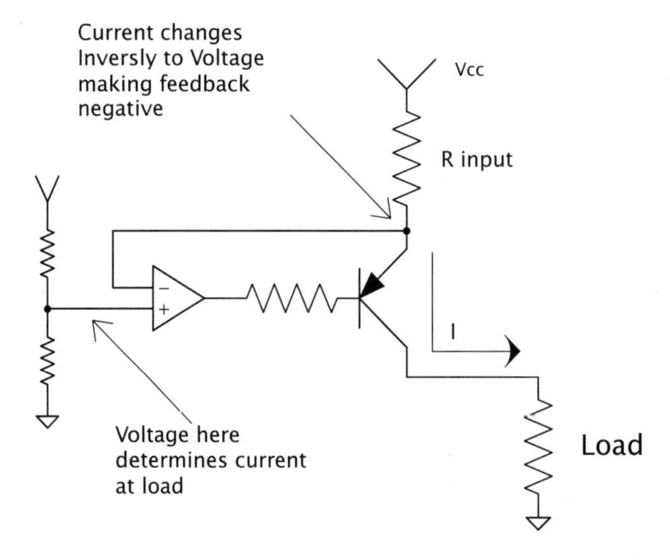

Figure 4-39 *Voltage-controlled constant current source*

Using negative feedback, the op-amp tries to maintain the voltage drop across *R* input. Even if the resistance of the load changes, the drop across *R* input stays the same. According to Ohm's Law, keeping *R* and *V* the same will keep current the same too. Remember, though, this current control has operational limits, it can only swing the output voltage so far to compensate for load variance. Once these limits are reached, the current regulation can no longer exist.

Get Your Own

These are just a few of my favorites. Get your own and know them well. You will be better served knowing a few circuit concepts inside and out than knowing thousands superficially.

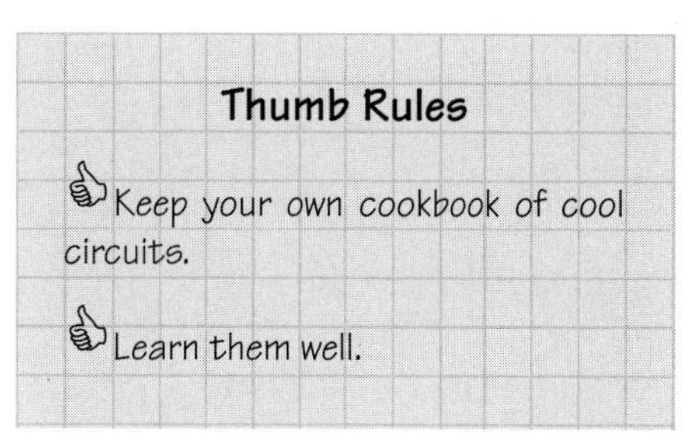

CHAPTER 5

Tools

What tools to use and when to use them? That is the proverbial question. There are innumerable tools available to the electrical engineer for use. This chapter is not intended to be an exhaustive analysis of various tools, but a guideline for selecting the proper tool and for setting it up right to get the information you are looking for.

Making the Invisible Visible

One difficulty that electrical engineers have in general is that the electrons they work with are not something you can pick up, touch or feel. (OK, I take back the feel part—there are definitely voltage levels that you can feel!) In reality, you infer their existence based on how they affect objects. You don't see the current flowing in a light bulb…you see the heat generated by the current flowing in a light bulb. Given this, tools that can measure various attributes of electricity are invaluable in electrical engineering.

Meters

Ahh, the lowly meter—probably the tool you will use most often in your quest for electrical engineering excellence. The first rule of thumb in using a meter (and this applies generally to all the tools you use) is to have some idea of what you are looking for. For example, if you are trying to read an AC signal, don't set your meter to DC. This may sound overly simplistic, but I believe poor tool setup is the most common mistake made by the average engineer. You can extrapolate from this rule to solve other misreading problems. This leads to a second rule and a case in point. The rule: don't trust auto set-ups implicitly. The case in point was a motor voltage we were trying to read; the voltage across the motor was a PWM signal with a peak of

140V. We were trying to read the average voltage across this motor with a Fluke 87, but the readings didn't make sense (note the application of rule one). We found that when the meter was in auto-range mode, the brains of the meter was confused by the PWM input. Setting the meter manually to the correct range resulted in an accurate and stable reading.

The two most common signals you will examine with a meter are voltage and current. In setting up a meter to read voltage, remember that you are hooking the leads up in parallel with the signal you are going to examine. When reading current, the meter must be hooked up in series in the circuit. Remember that current is a measurement of flow. Nearly all meters require you to hook the leads into different plugs when reading voltage vs. current. This is so the signal can be routed through an internal shunt resistor across which a voltage is measured and scaled to represent current.

Typically, there is a fuse in the meter to protect this shunt from overload. On some meters the shunt is a different value for different ranges of current.

All meters will affect the circuit they are hooked to whether they are in voltage mode or current mode. The question you should ask is "How much?" A typical DMM (Digital Multimeter) has 1M to 10M of impedance in the voltage-measuring circuit. As you hook the leads up to the circuit, consider that you are adding a resistor to the same points.

Let's look at an example. We will assume our meter has a 10-MΩ input impedance and we are measuring the output of a voltage-divider circuit.

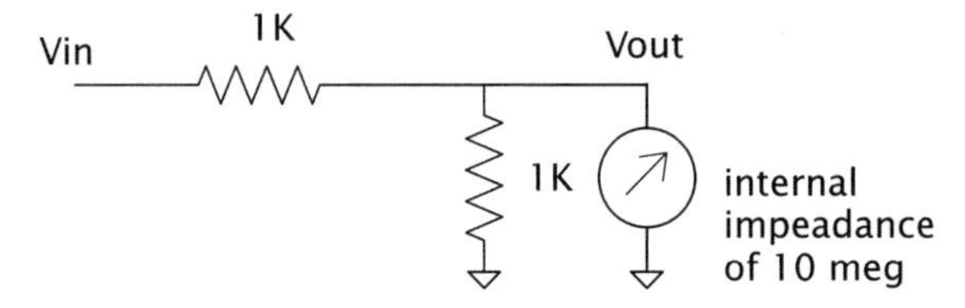

Figure 5-1 *Equivalent circuit of a meter on a 1K voltage divider*

equivalent circuit

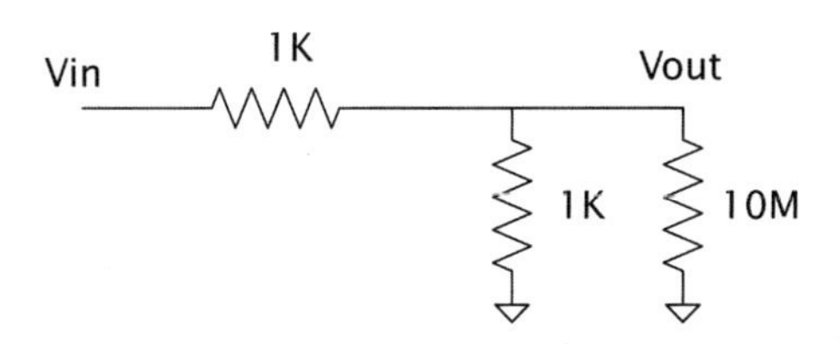

Let's calculate the effect this has on the circuit. We will start by calculating the parallel resistance of the meter and the resistor it is hooked to.

> Hmmm, Scribble scribble, nibble on the pencil eraser, mumble to myself, I just learned that rule, it's the product over the sums, so that would be (1K * 10M)/(1K + 10M) or 0.9999K.

Now we apply the voltage divider rule. More humming, more scribbling, and we see that without the meter the output will be at 2.5V, but with the meter the output will be 2.4999V. So we will probably all agree that the meter does not have a significant effect in this case.

Let's change the value of the resistors and see what happens. We will make them 1MΩ resistors.

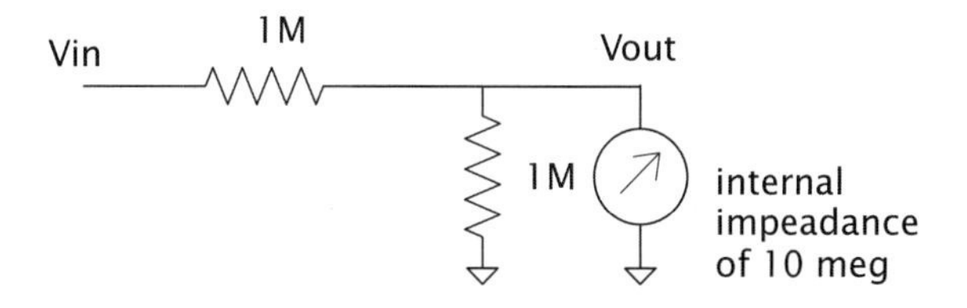

equivalent circuit

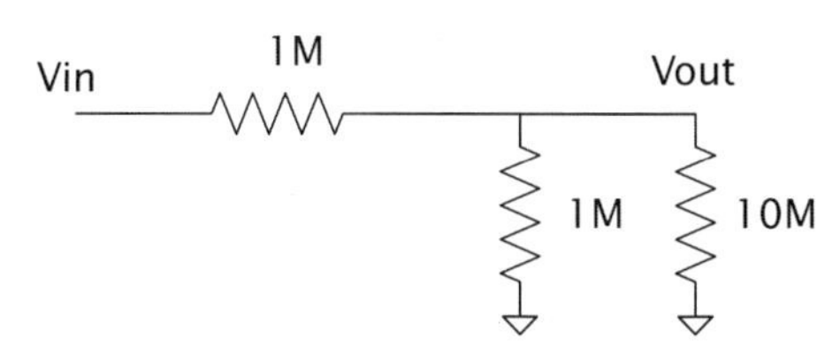

Figure 5-2 *Same meter on a 1M voltage divider*

The first thing you should notice is that without the meter the voltage output will be the same as the previous circuit. But what happens when you hook up the meter? 1M//10M (the "//" marks mean "in parallel with") gives a value of 909.09K. Run that through the voltage divider rule and you get 2.3809V at the output. Do you see how the meter can make a difference? Hopefully, what your intuition is telling you is that the effect of the meter depends on the ratio of the meter impedance to the impedance of the circuit you are reading. Now try an experiment. Change either resistor in the 1M divider to 1K and run through the same analysis. You will see that the meter no longer has a significant effect. This is because the overall impedance of the circuit is about 1K. Thevenin taught us that. If you don't quite follow, now is a good time to flip back to Chapter 2 and bone up on Thevenizing. Make sure you consider the overall impedance of the circuit you are measuring when determining the effect a meter will have on your circuit.

Scopes

The primary two controls on a scope are just like in Outer Limits: "We control the vertical and the horizontal…"[1]

In other words, on a scope you are controlling the voltage per division and the time per division. The divisions referred to are the vertical and horizontal marks that make a checkerboard on the screen.

The next most important control is the capture mode, whether you are seeing a DC or an AC signal. Unfortunately, this control is usually somewhat hidden. This control is important because it can affect the way a signal looks on the screen. (Just take a 0 to 5V logic signal and read it with your scope in AC mode and you will see what I mean.) In AC mode, the inputs are connected via a series capacitor to the guts of the scope. This removes any DC offset the signal might have. In DC mode, the voltage level of the signal relative to the ground lead of the scope is maintained.

The oscilloscope is, in my opinion, the single most useful tool an electrical engineer can have. That said, I've seen a lot of engineers chase down blind alleys because they misread their scope. Correlating these two facts indicate that it is very important to know how to set up your scope.

First a caution: never trust the auto set-up on a scope. Let me repeat—never trust the auto set-up on a scope. Make sure you know what you are looking for. This is even more important than auto set-ups on meters because of what the scope might do.

---

[1] It is funnier if you think it in the same deadpan voice that the old TV show used. For those engineers who are too young to have any idea what I am talking about, you'd better Google Outer Limits.

For example, say you want to measure a 5V signal that switches to ground when you press a button. You hook up the scope, press auto set, and then press the button. The most likely scenario in this case is the scope sees a 5V DC signal and starts hunting for some frequency to look at. So it zooms in until you see a 10 mV AC ripple from the power supply at 60 Hz. Now you have a scope set to 10 mV per division vertically and 10 ms per division horizontally in AC mode. Remember, you were trying to see a 5V DC switch to 0 when you press a button. The auto set totally missed what you were looking for. You probably won't even see the switch action at this setting and, to top it off, there will be a 60-Hz ripple on the screen to confuse you.

This is the most common mistake I have seen. An engineer hooks up a scope to the misbehaving circuit, hits auto setup, the scope zooms in on an irrelevant signal, the engineer, thinking "Aha, I have found the glitch!" spends the rest of the day chasing something that doesn't matter.

Having an idea of what you are looking for is an equally important rule of setting up a scope. Ask yourself how long the signal will last. What voltage levels do you expect? Start with those settings on your scope. Now, once you are capturing what you expect, zoom in on the details to look for those pesky glitches. Say, for example, you suspect a switch bounce on the example we used above. Start by capturing the signal 5V and 500 ms per division. After all, you are pressing this button—just how fast are you? Once you can reliably catch this signal, start working your way in; go to 2V or maybe 1V per division to increase vertical resolution. Then start working on the time base. Decrease the time per division while periodically checking the signal you are watching. This way you drive the scope to look at the signal you want to see. If you let the scope do the set-up, it is kind of like being kidnapped and driven around blindfolded. When you take the blindfold off, you don't know where you are. You will be lost, confused and disoriented and that can lead to wrong assumptions. If you are the driver, on the other hand, you know how you got there and have a better idea of what is going on.

So set-up is important. Here are some other general things you should know.

Ask yourself, "Is the signal really there?" Why? Because it is possible that the scope with its high impedance is picking up noise that really isn't affecting what you are looking for. Try this: disconnect the leads. Is the signal still there? If it is, that is a good sign you are dealing with a radiated noise that may not even affect what you are looking at. If you are working with high-power circuits and switch-mode supplies, there will be all sorts of artifacts that really don't affect anything, but pick up nicely on the antenna of a scope lead.

Make sure you hook up all your ground leads (even though on most scopes they are tied together internally). The reason to do this is because small currents flowing back through your scope ground can lead to incorrect results. You may even think you have discovered free energy.[2]

On most scopes the ground lead is connected to the earth ground of the scope (for safety reasons), which can be disastrous when looking at certain signals that may reference to a different point. You can get currents through the ground leg that throw off your reading at best and blow stuff up at worst. If this is happening, get an isolated scope.

Just like a meter, high-impedance circuits can be affected by the scope leads. Have you ever had a problem go away as soon as you clipped the scope on? Try a 10-Meg resistor or 100 pf cap across the same connections. It is a good bet that will fix the problem (in case you were wondering about where those values come from, they approximate the impedance of a scope lead).

When all else fails, swallow your pride and read the manual. Yes, I know it's hard, but the destructions[3] usually give you insight into setting up the scope so you see what you want.

Scopes these days have myriad features, cool glitch captures, colored screens (a personal favorite of mine), magnifications, yes, auto set-ups, and much more. The point here is to get the basic set-up right so when you use those other features, you have an idea of what is going on. Remember, getting what you want out of the scope is up to you, at least until they get that mind-reading function working…

---

[2] This is a whole other topic for a whole other book.

[3] Or instructions, depending on how you look at it.

Logic Analyzers

A logic analyzer is similar to an oscilloscope in that it displays a signal over a time base. It differs in two main aspects, the first is it displays only logic levels; the second is it has many more channels.

Think of a logic analyzer as a digital-only oscilloscope. It is not going to show you signals between a logic high or low. There are logic analyzers with a couple of scope channels built in to get around this limitation, but if you don't have one of those, make sure you understand you are seeing the logic level closest to the signal you are reading. If the level the analyzer considers a high or low differs from the level your circuit does, this could lead to confusion. If you suspect the logic signals are not reaching the required voltages, make sure you check it with a scope.

The best feature of a logic analyzer is the fact that it has so many channels. This becomes very useful when you are trying to observe all eight or more lines on a data bus at the same time. It's kinda hard to look at eight things at once with only a couple of channels.

This, like all the others, is easy to set-up wrong if you have no idea what you are looking for. Don't just set it up blindly—have an idea of the time base needed to find what you are looking for. Also, remember that it is designed to display logic signals, possibly masking signal levels that you may not expect.

These days, with their digital storage capabilities, scopes are closer than ever before to logic analyzers, and the fact that many analyzers have some scope-like capabilities make them more scope-like than their predecessors. If forced to categorize, I would say that a scope is a more general tool that can be applied in nearly any situation except the one where you need to see a whole bunch of channels at once, and in that case the logic analyzer is definitely the tool of choice.

Remembering the basic rule of thumb with this tool, as with all others, is to have an idea of what you are looking for. If you do so, you will find this an effective tool to have at your disposal.

Thumb Rules

- Always have an idea of what you are looking for.
- Don't trust auto set-ups.
- Don't trust auto set-ups.
- Is the signal really there? Unhook the leads and see if you still pick it up.
- Hook up all the ground leads.
- The higher the impedance of the circuit, the easier it is to disturb with measuring tools.
- Read the manual!
- And one last time, don't trust auto set-ups.

Simulators

First, let me make a statement. Simulators are great tools, (here it comes) BUT too often I see a major mistake made with a simulator. The engineer fires up the simulator, tries out his or her idea, gets it all designed, then proceeds to build a real circuit, only to find the circuit does not work as the simulation did. Here is where the mistake comes in. All too often the engineer spends all his time trying to figure out why the circuit isn't working right, implicitly trusting the simulator to be right. For some reason as soon as the circuit is modeled on a computer, it seems to be an engineer's nature to trust the result they get on the simulator without question. Doing so almost invariably leads to immense frustration and confusion. You should take this adage to heart. *The real world isn't wrong, your simulation is.* It is always true. If the results don't match, something in your simulation does not actually represent what is on the prototype in the lab. The simulation is a representation of the real world, not the other way around.

What is Real?

This is not to say that the circuit on the bench is what you want it to be. It very well could have a mistake in it that is not in your simulation. But that doesn't change the fact that the simulation is not truly modeling your design. I have found if you take the perspective of always questioning the simulation, two things happen. First, you gain an intuitive understanding of how different components affect your circuit. As you fiddle with the simulation, trying to get it to match the real world, you begin to grasp how large of an effect this or that component has. Second, you learn about the limitations of components in the real world—something that just studying math and formulas will not give you. Take, for example, a 10-µf electrolytic capacitor in this circuit

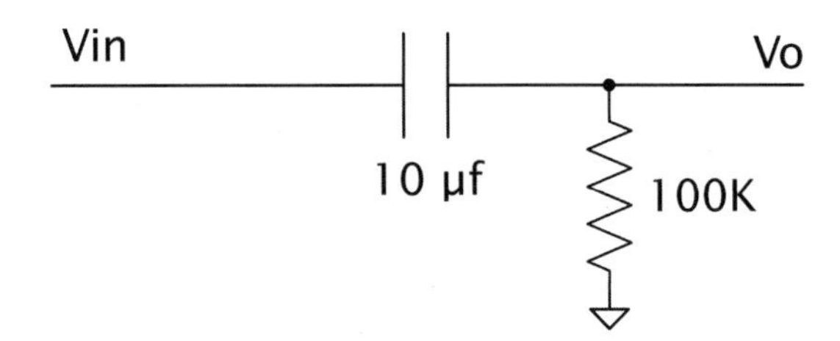

Figure 5-3 *RC high-pass filter*

According to all the formulas you have learned, this should pass all the high frequencies above 1/RC you would ever want. Just about every simulator you find will do so, but hook this circuit up to a signal generator and you will find that, as you get up to the higher frequencies, it doesn't work as well as the math says it should. The math isn't wrong, the component just isn't perfect. Some simulators will allow you to create equivalent circuits to more accurately represent a given component. The problem is that doesn't negate the need for you as an engineer to understand the limitations of the components. You really need to have an idea of what is going on or the simulation can lead you down a fruitless path. The skill of estimation is immensely important when using a simulator. Skip back to Chapter 1 if you need to brush up on your hand grenade skills.

A Powerful Tool

Now that I just got done bashing simulators for not dealing well with imperfect components, let me say that, ironically, they are potentially the best tool you have to create a design that handles imperfect components well.

Once you truly understand the variability that can occur in the parts you are using and create an accurate model of what they do, you can do something with a simulator that you cannot do easily with actual parts. You can build thousands of pieces of your design in cyberspace with each part varying a little from their nominal values. You can swing the tolerances to their extremes with the click of a mouse, saving a hunt through a drawer for that part that is on the low end of spec. If used correctly, a simulator is probably the best tool you have to make your design handle the inherent variability in components.

Develop Your Intuition

One of the best things you can do with a simulator is to use it to develop your intuitive understanding of basic components. Every engineer should simulate the transient response of the basic RC, RL, and RLC circuit. Try changing the values of the parts just to see what happens.

If you start modeling simple circuits and getting confidence in making the model accurate, you will be much more successful as you create more complex simulations. It's not unlike learning to play the guitar, you don't just sit down and rip out a lick Eddy Van Halen would be proud of. You need to be able to handle the basic cords first. You should learn to "play" a simulator the same way. Even though it is easy, don't put together your whole design in the simulator the first time and press go. If you do, I can nearly guarantee you will get confused by the results and they will probably be wrong as well. Break your circuit down into simpler pieces, ones that you can intuitively understand and simulate those parts first. Eat the elephant one bite at a time.[4] When you are sure your model represents the real world accurately enough[5] for the problem at hand start knitting those pieces together and see what happens.

One word of warning: playing around with a simulator can be very time-consuming. Don't get so caught up in doing the simulation that you never get around to building an actual circuit. In fact if you are unsure as to how the circuit will really work, go build it up in the lab and see. When it comes to tolerance analysis, you should already have a real circuit in the lab running when you start simulating. Get the circuit working with nominal values before you start investigating what component variance will do. Simulation should go hand in hand with lab work.

---

[4] See Chapter 1 way back at the beginning for the elephant reference.

[5] Remember that accuracy is relative. If you don't need to know the answer to four decimal places, don't waste time trying to get that close.

Thumb Rules

- The real world isn't wrong, your simulation is.
- Gain confidence in your model representing your design accurately.
- Use estimation to double check your simulation (a couple of more 'tion's and this could be fun to say!).
- Model basic circuits to develop your intuitive understanding of the basic components.
- Break the model down into pieces that are simple enough to check for accuracy. Then add the models together.
- Simulation goes hand in hand with lab work.
- When setting up you tools, have an idea of what you are looking for. How fast is the signal? What voltage level do you expect it to be at? Etc., etc.

Soldering Irons

I was passing by the lab one day when I saw one of my technicians looking over the shoulder of one of the engineers who was doing a less than spectacular job of soldering components on a PCB. He had but one comment. He said, "What we have here is an engineer trying to do a technician's job." Then he sat down and proceeded to do a most excellent job putting the board together.

On the chance that you may not have a skilled tech at your disposal, and the fact that I believe the more you know about how the product you are designing goes together, the better designer you will be, we will go over the basics of soldering.

The Basic 4

Making good solder joints requires four things: cleanliness, solder, flux and heat.

First the parts need to be clean and dry. If the pads are corroded, often a little rubbing alcohol will clean them nicely.

Second, you need solder. Solder is a mixture of lead and tin with a melting point around 100° to 200°C depending on the alloy used. When applied properly, it will provide an electrical and mechanical connection between the part and the PCB. Although it is a mechanical connection, remember it is not a particularly strong mechanical connection.

Third, you need flux. When hand soldering, this is often inside the solder wire in the hollow core. What is flux, you ask? Flux is a chemical that cleans when you heat it up, preparing the joint so that the solder will stick well. In some cases the flux is applied before the solder, such as before it goes over a solder wave or into a solder bath. Flux is also called resin.

Last, you need heat. Heat brings it all together. The solder will flow to where the heat is. This means you need to get the leads of the part heated to make sure the solder flows. When prototyping, the typical way you get heat to the part is with a soldering iron. Some other ways are hot air pencils and reflow ovens, but the same thing applies. Heat makes the solder adhere to the pad and the lead of the part. When you are all said and done a good solder joint looks like this.

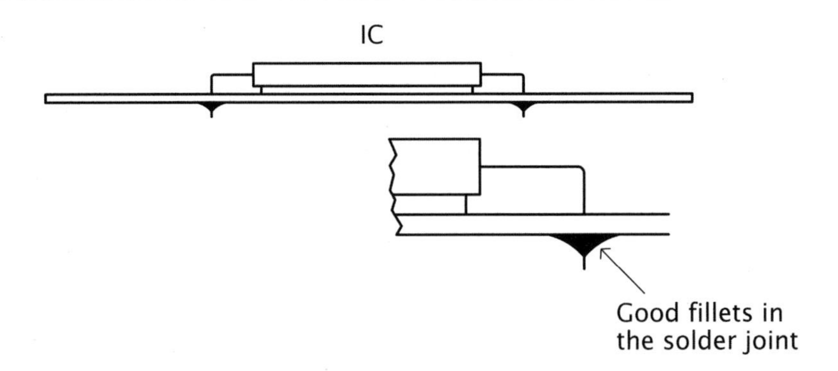

Figure 5-4 *Good solder joint*

Solder Goobers

Of these four items, the one that usually causes problems is the application of heat, particularly when you are using a soldering iron. Parts and PCBs are both sensitive to heat. The parts can be damaged by too much heat, and the PCB pads are adhered to the PCB with glue that has a lower melting point than solder.[6] Too much heat for too long can be bad. Parts can be damaged and pads or traces can be lifted (when the glue is melted).

The flip side is that not enough heat will lead to failures. One of these failures is called the cold solder joint. This happens when you do not get enough heat to both parts being joined. When this happens solder will adhere to one part and not the other. The part that did not get enough heat will not get a good connection. That is why it is said to be a cold joint. It looks like the following figure.

[6] It is actually intended to be this way because during soldering the copper traces will expand (due to heat) at a different rate than the PCB substrate. If the glue is melted, this keeps the trace from deforming.

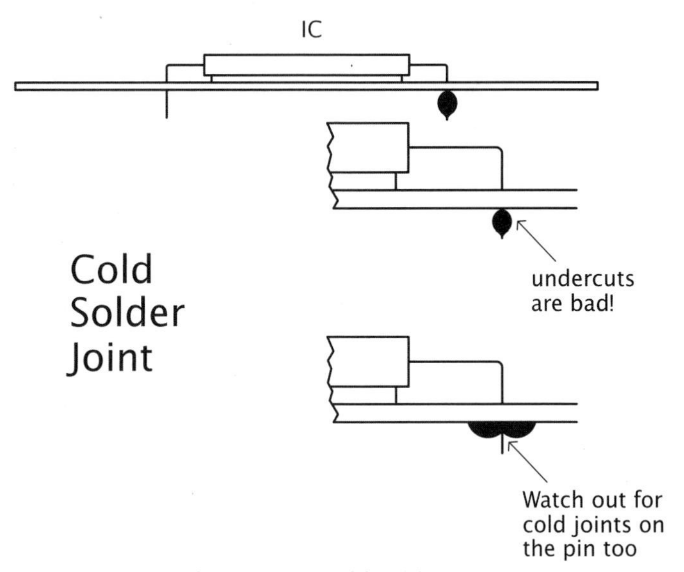

Figure 5-5 *Cold solder joint*

A cold solder joint is the most common failure made when using a soldering iron. You get going a bit too fast and don't leave heat on the joint long enough, or only touch the iron to the pad and don't get it on the lead. A good rule of thumb when soldering by hand is to place the tip of the iron on the joint, count "one Mississippi" and then apply the solder, wait a moment and remove the iron.

There are two other things you need to do to keep your soldering iron working right. One is to make sure the tip is tinned. If an iron is left on for some time, the solder and resin on the tip will evaporate, leaving a dry tip. A dry tip will not conduct heat to the parts you touch as well as a tip with solder on it. Applying a little solder to the tip before using it is called tinning. (You can also tin wires to make them easier to solder to a connection.) If you are having a problem getting heat to a part, try adding a little solder to help conduct the heat.

The second thing you need to do is clean the tip of the iron often. Any decent soldering iron will have a sponge in a tray with water. Wiping the tip on it will effectively clean it. Cleaning the tip keeps the build-up of excess flux from interfering with the soldering process. A word of caution: don't soak the sponge with too much water, and don't rub the iron on the sponge excessively. Too much water or rubbing it too long will cause the tip to cool down too much, affecting the next joint you need to apply solder to. Don't forget to tin the tip before going onto the next joint.

De-Soldering

Unless you never make a mistake, at some time in your career you will need to remove a part that has been soldered to a PCB. De-soldering can be a frustrating experience. It is during de-soldering that you are most likely to lift a pad from a PCB, burn your fingers, and possibly cut loose with a few expletives. Hopefully, I can share some hints to keep the air from turning blue when de-soldering is required of you. We will also discuss the three main tools for lifting solder: solder tape, hand pumps, and de-soldering stations.

Hint #1

Sacrifice the component if possible. If you do not have to salvage the part you are taking off, clipping the leads (so you are not trying to remove a 40-pin part all at once) is a great help. Cut all the leads and deal with one pin at a time. Once though we had a situation where we needed the 40-pin part, but not the PCB. Our solution? Take the board down to the shops and hit the back of the PCB with a quick burst from a blow torch while yanking the part off of the other side with a pair of pliers. Worked like a charm and the burnt PCB made for a great joke on management later.

Hint #2

Add solder to the part. Adding solder can help you conduct heat to the joint you are trying to dismantle. The trick to getting the part off is to get heat quickly to all the places you need to. For example, you might need to remove a radial electrolytic capacitor. On this part both leads are close together. You can actually create a solder bridge between the leads and get heat to both leads at the same time and quickly pull the part off.

Hint #3

Get the part and pin off before you worry about getting the solder off. Apply heat, yank the part, then come back and get solder out of the hole. Often when you are trying to get the solder completely off before taking the part off, you will find that a small piece of solder still holds the lead to the side of the via. Trouble is it is such a small piece of solder that it is difficult to heat it up to get the lead loose. Apply hint #2 and try again.

Tape

Solder tape is a copper braid. Copper, being a great conductor of heat, will wick the solder into the braid when heated up. It is important to apply the heat to the braid and then press the braid on the solder. If you just try to stick the braid in the molten solder without heating it up, the solder will just sit there. Remember, solder flows to heat.

Also note that the braid is made of copper, and copper can tarnish. Once it has tarnished, solder will have a hard time sticking to it so old solder tape is pretty much useless. New tape though works well and is cheap and convenient to use.

Hand Pump

My own tool of choice, the hand pump, is easy to use, relatively inexpensive and easy to maintain. When using the hand pump, you press down the plunger, heat up the solder you are trying to remove, press the button and thwoop, in goes the solder like a spaghetti noodle. Make sure you leave the iron on long enough for solder to become molten clear through the via. You may need to apply hint #2 to help things out. The biggest down side to the solder pump is the sore thumb you are going to get if you need to do a lot of de-soldering.

De-Soldering Station

If you need to do a lot of de-soldering and you have some cash to spend, this is a tool you need. The de-soldering station is a powered version of the hand pump. The iron is integrated into the tip where a vacuum is applied to suck out the solder. Generally you need to maintain these tools regularly. The tips can wear dur to the corrosive nature of the solder removal. They can get plugged easily when not used properly. Always suck to the side, not straight up. The molten solder has weight and trying to move that to the side is easier than trying to lift it straight up. Keep sucking for a couple of seconds after the joint is clear to make sure the molten solder gets all the way into the receptacle in the gun so that it doesn't solidify midway through the nozzle.

Properly maintained, this is the quickest and easiest way to get solder off of a PCB. It is also possible to get a part off with the pin still in place. This is done by using a small circular motion to get the pin out of contact with the via as you are sucking the solder. However, it is still easier to sacrifice the part if that is possible.

Thumb Rules

- Solder goes where the heat is.
- Solder goes where the heat is.
- And if you didn't get it this time, remember solder goes where the heat is!
- Prevent cold solder joints by counting one second while applying heat.
- Make sure you tin the iron before using it.
- Clean the tip often.
- When de-soldering sacrifice the part if possible.
- Add solder to promote heat flow.
- Get the part and pin off before worrying about getting the solder out of the hole.
- A small circular motion with a de-soldering station tip will help clear the solder from the lead and the via.

People Tools

When I entered the professional realm for the first time I had an experience that I still remember. I got this call from the receptionist. She said, "So and so is here to see you. He wants to know if you can have lunch with him." Of course I'm thinking hey, free food, but who is this guy who seems to be my instant friend? Thus it was over nacho chips and arroz con pollo that I tumbled headlong into my first experience in the world of reps, distributors and FAEs.

Lunch was good. I had no problem figuring out what to order from the menu, but getting to understand the roles of these three people took more than a few tacos. It can be a bit confusing as to who does what and what that means to the average "Dilbert" out there, so I figured it wouldn't hurt to give you some idea of what these guys do and how they can help you.

First of all, all these guys have some relationship to the company that makes the product you need, be it an IC, transistor, micro or whatever. When I say company in this case I am referring to the company with the product to sell, not the company you work for.

The Company

The company selling the widget you are interested in employs several layers of people to get their product in front of you and sold to you. They also have internal sales persons and managers that you may get to know if you work closely with them. How close you work with them often depends on the amount of business or potential business you represent for them.

If you can get to know someone on the inside, it is never a bad thing. These guys are more accessible now than ever, and if you need to know how a part will act in some weird situation, talking to the guy who actually designed the part is definitely the best option.

The Rep

One layer removed from the company you will find the rep. This guy is the person that represents the company making the part they want to sell to you. He does not usually draw a salary directly from the company. He is paid by the rep firm that represents the company. He and/or his firm are typically rewarded for his efforts with some percentage of the sales he makes, usually between 1% to 5% depending on volumes and other complicated formulas designed to cost the company making the product the least amount of money, yet drive sales as much as possible.

The rep will work with the distributor scheduling parts, getting you samples and other such stuff. He is very interested in you using the company's widget and it is unlikely he will offer an alternate solution since he is monetarily tied to the company's widget. Reps typically are not allowed to represent competing firms. My experience is that reps for the Dilbert-esque products we are talking about have an engineering background of some type.

The Distributor

"What is the difference between the distributor and the rep?" I asked one of these guys once. "About 15%," I was told.

Distributors will stock parts and mark up the cost to cover the money they have expended. They tend to make about 20% on a given part, but that is just a ballpark figure. The actual number can be all over the place, depending on the particular business agreements in place. Some manufacturers force distributors to specific margins if they carry their parts.

One of the biggest distributors out there watches market trends constantly and looks to buy stuff that is likely to become rare yet needed in the future.[7] Then they go out and buy a whole bunch of said item, sit on them for a bit, and sell them at a profit later.

[7] Slight pun intended, if you get it, I don't need to explain it, and if you don't it is no big deal.

Sometimes companies use exclusive distributors. Some use multiple channels. In the case of multiple distributors, whoever is the first to register a part for a particular application gets a lower price on the part than any of the other distributors. This is designed to reward the distributors for getting out and getting more business.

The biggest advantage of a distributor is supply chain management. By buffering stock for you they can help handle ups and downs in order sizes, shortening lead times when orders go up unexpectedly.

They are less likely to be tied to a particular manufacturer of a part, and often carry multiple solutions to a given problem, they will tend to lead you to the part that will solve your problem and be the most profitable for them.

Distributors are less likely to have an engineering background. Hopefully many of them will buy this book and not be upset that I disclosed so much about this seemingly secret world.

FAE

Working for distributors, reps, or even the company you will sooner or later run into the field application engineer (FAE). The FAE plays several roles. First he is the main guy who helps you get the part to work. He or she also looks at your application and often might suggest parts that would be a good fit. Lastly, often they act as a translator between you and the distributor. As you may well know, it in can be difficult to understand an EE when they get into technical details.

For many a Dilbert, the FAE has the perfect job. They get to come up with all these solutions, but have no responsibility for actually making it work at the end of the day. There have been days I have dreamed of being an FAE for just that reason. However, the flip side is they also rarely see the finished product and miss out on the satisfaction of the "being late and over budget, but whew! It's finally done" feeling.

FAEs often go to a lot of training sessions with the company to understand how the part works. They usually know or can contact the engineers in the company to help answer questions.

More and more in an effort to sell their parts, companies are not only developing new parts but also creating applications for those parts. FAEs and company engineers are often tasked to come up with cool little application demos and the like that show you how great the part is. Remember, though, for them it is like lab back in school—they only need it to work that once when they are showing it. Production runs can be a whole other matter, so do your homework and test an FAE design thoroughly before you commit it to full on production runs.

Design Wins

This is a common term to hear over the appetizer. When the distributor registers a part with the rep and thus the company, and then the part actually gets used in the design, it is called a *design win*. Odd how engineer types use words that make sense when you think about it, isn't it?

Remember, whoever registers the parts gets a discount on that item. That usually makes their price hard to beat compared to other distributors. I say usually, because I have seen a lower price quoted from an unregistered distributor, albeit rarely.

Another thing that can happen is registration can be moved. If you really don't like working with the distributor and you are a big enough customer to the company, they can move the registration to a distributor you prefer. However, this is rare and usually done as a last resort because if it happened too much, all incentive to get their part in the door first dries up.

Going Direct

Depending on the size of your orders, one thing that you may consider is going direct. This means you will buy parts directly from the company, skipping distribution. The goal is to get a lower price.

The cons to this are several. The company will usually have minimum orders, lead-times and terms that are less favorable than working with a distributor, and if you are a little guy (order wise) they probably won't even consider it.

Before you do this, consider the options carefully, because you will be removing a piece of the support structure that you use in the design and supply management of the part.

There are companies that will not even allow you to go direct; they have a policy of distribution only, but I happen to know that they also dictate to those distributors what the final price will be to keep them competitive in the market place.

To Sum It Up

There are several legs to the stool of getting parts to you. Each one wants his piece of the pie and has services to provide to justify their cut. Knowing who does what will enable you to better use these people tools to succeed at your job.

In my experience, the more successful reps, distributors, FAEs and the like will visit often enough to know what you are working on and keep you in mind as they see new technologies and ideas that you can use. They will have suggestions and solutions and yes, they may even buy you lunch once in a while.

Thumb Rules

- The rep works for the company under a contract.
- The distributor works for himself.
- The FAE knows how the stuff works.
- The company wants to sell you a cool widget.
- Sometimes you can get the company and the FAE to help do some of the design work.
- All of these guys can help you find parts and get quotes; they work together to provide the best service they can.

CHAPTER 6

Troubleshooting

The perfect design approach doesn't guarantee everything will go flawlessly in production. One engineer I work with is fond of saying, "I'm not happy till ten thousand pieces have gone down the line OK." I've seen tolerance problems appear in production after half a million pieces have been run. It is very difficult to predict and prevent something like that.

The fact is, the more thoroughly you try to analyze the design, the longer you will be waiting to produce it, but if a product never gets into production, no one gets paid (except for government work of course). So a balance needs to be struck between design analysis, testing, and production runs. This being the case, you will likely be faced with trying to determine a production problem at some point in your career. How will you go about it? What approach will you take? Hopefully this chapter will give you some ideas.

Getting Ready for the Hunt

As we discuss the topic of finding trouble and shooting it, I will refer a lot to my own experiences. I am sure you will have unique and often completely different results. The idea here isn't to tell you what the problem is in your design, but to give you some guidelines you can use to troubleshoot the problem yourself.

Shotgun Wedding

We will get into the interesting title a bit later. While it may seem a little out of order, I think we should cover one item first because it is so important to the rest of the process.

The first rule of thumb is: Don't discount a theory (no matter how obvious or ridiculous it may seem). Try to prove it right or wrong by experiment and then move on to the next idea. Too often you will be carrying an assumption that you won't even realize that can lead you to a wrong conclusion. It is vitally important that you have a process to check and validate a theory. Without that, you will be forever jumping from one idea to another without ever coming to a conclusion. With that said, let's move on.

When it comes to troubleshooting methods, I group them into two common categories:

> *Scientific method*: Do what any good detective would do—look for all the clues you have been given and deduce what the problem might be based on experience and knowledge.
>
> > Advantage: Eventually you will identify the problem.
> >
> > Disadvantage: It takes a lot of patience and time.
>
> *Shotgun method*: Take a shot at as many possibilities as you can and hope you get a hit. Sometimes you get lucky and you solve the problem fast.
>
> > Advantage: If you are lucky, you will solve the problem fast.
> >
> > Disadvantage: If you are not lucky, you will chase around in circles forever.

While both these methods have their place, what I propose so subtly in my title is a third approach to troubleshooting. I call it the Scientific Shotgun Method—a marriage if you will, of the shotgun method and the scientific method. Start like this.

When a problem first comes to your attention, take a shot at as many possibilities as you can. Write down all the things you think might be causing it. Use your intuition as well as your experience in this exercise. Speaking metaphorically, get out the shotgun, take aim and fire. Then let the scientific method kick in, and figure out a way to evaluate each of your possibilities to prove or disprove them, and have at it.

When employing the scientific shotgun method, based on my experience, results like these are typical: 7 out of 10 times it will be something stupid that the shotgun method catches easily and quickly. For example, using an old software version, or a component wasn't stuffed, or a fuse was burned out, etc. An average of 2 out of 10 times, something more subtle will be found that takes some trial and error and requires new data to be found and evaluated till the problem is solved. About 1 out of 10 times the solution takes a longer time, but eventually is found by repetitive applications of both methods, where the shotgun approach opened up new areas of research that scientifically lead to the resolution. On the aggregate, problems are typically solved quickly with a minimum of running in circles when the *scientific shotgun* approach is used. (Did you ever think you would see those two words together as something meaningful?) This is a real boon in a consumer product world when shipping that new design on time is all-important.

You Too Can Learn to Shoot Trouble

Have you ever seen an engineer having immense difficulty in diagnosing the cause of a problem, when a lowly tech stops by and identifies the bad part right away? Or maybe you've seen a tech struggle for days only to have the engineer take one look at the schematic and say, "There is your problem."

Some people have trouble with troubleshooting, and others just seem to have a knack for it. If you ask them to explain what they do to solve it so quickly, they are often at a loss as to how they do it—they just do. Believing that you can learn anything even if it doesn't come naturally, I have distilled down into thumb rules some of the things that those guys with the knack do.

Simple Things First

After you have made the list of things that could go wrong, start with the simple things first. My father recounted an experience to me when I was younger that really stuck with me. He basically rewired an entire car looking for an electrical problem. To his dismay it turned out to be a bad fuse. Looking at it, it appeared OK but when measured, it was open. This may seem like a "duh" moment to the outside observer, but it is an easy trap to fall into. The way to avoid it is to check out the simple things first. Does the chip have power at its pins? (Not just to the board.) Is the oscillator running? And so on.

Look Outside Your Specialty

It's hard to make a blanket statement about what is likely to fail, as often there are many small clues to a particular problem. To further complicate things, it is often a combination of more than one factor causing the problem.

It is human nature to focus on what you know; everything else seems somewhat magical after all. Good troubleshooters are often good generalists. They know a little bit about everything and use that to connect the cause to the effect. They always want to know why this is that and what does that thing do, and so on.[1]

Sometimes there may be seemingly insignificant clues. One time early in my career we had a problem with some displays we were producing. A percentage of them were failing and I was assigned to find out why. When I took the unit apart, it would function correctly. When I put it back together, it would fail again. I looked for hours trying to find problems with pinched wires and cold solder joints, to no avail. So I sat there and stared at the PCB for a while. As I did, I noticed two small marks on a resistor; I wondered where they had come from. After some examination, I discovered a screw head that would contact this resistor when the PCB was installed. It turned out that the screw head would short across the resistor when the PCB was installed, making the part fail. When I removed the screw, the console worked correctly after assembly. Don't be afraid to look outside what you know for the cause of the problem.

Don't Ignore Anything

Try to keep track of all the clues to a particular problem. Keep a list of symptoms and clues that you can refer to in your deductions. Don't ignore anything, as one fact may connect with another to point you in the right direction. Here is a case in point.

During testing of a circuit my engineering team had designed, we had been experiencing random unexplainable problems. The test engineer made the statement that these problems seemed to have started when we began using surface-mount PCB designs. We were completely at a loss as to any connection between this and the problems. Then I remembered, when looking at one of the circuit boards, I had

[1] You know—the kids that moms can't stand because they are always asking questions. I know the personality type well. I have five of them!

noticed some small black fibers that appeared to have dusted the PCB. The test engineer initially dismissed this at first as small bits of plastic that accumulated in the environment of the circuit during use(this made sense as there was a moving belt made of plastic that could leave these bits as it wore down). He was sure it wouldn't make a difference.

However, we knew there were points on the PCB that, if they were shorted by even a few megaohms, could make the circuit repeat the problem that we were seeing. Connecting that with the fact that the surface-mount components would have closer spacings made such a short more likely. I insisted that we determine conclusively whether or not these fibers were conductive. The first thing we did was collect a sample of this "dust" and bring it near a magnet (on the presumption that if it is ferrous it is likely conductive). We were surprised at how much ferrous material was in these presumed plastic shavings. It reminded me of the classic physics experiment where you put metal filings on a piece of paper and then move a magnet underneath to see the field interactions. Once we protected the board from this contamination, the strange behavior stopped.

By not dismissing the obvious presence of these fibers, combined with the clue that it started when we went to an SMT design, we were able to make a connection that allowed us to solve the problem.

Which One of These Things Is Not Like the Other

Did you watch Sesame Street as a kid? One of my favorite segments was which one of these things is not like the other. You were taught to identify similarities and then point out the one that just didn't seem to fit. This is a very important troubleshooting skill. All the good skills work in more than just the "world of sparkies." They can be applied in just about any problem hunt. Here is another case from the archives of Darren.

Years ago our fridge stopped dispensing water. I figured I should just tear into it and take a look. After all, the water valve was controlled by a solenoid. That was close enough to electricity for me. There were two valves, one for water and one for ice. I tore these valves apart and notice some wear on a rubber washer inside the water valve. The solenoid pressed this washer against a hole in the valve. Little bits and pieces of rubber were falling off, it was so worn. This became especially important after I looked at the ice solenoid (it was operating correctly) and the rubber washer on that one didn't show wear.

It just didn't seem right for the washer to be falling apart like that. It just didn't fit. So I replaced the rubber washer. Put it all back together and, voila, it worked great. The skill in this case was looking for something that just didn't seem right. Sometimes you can figure this out by asking yourself, "Would I have designed the washer to fall apart?" The obvious answer in this case was no, so hence something was wrong with the washer.

Estimation Revisited

Sometimes it seems we spend half of our time designing a circuit, and the other half trying to figure out why it isn't doing what we designed it to do.

Back in Chapter 1 we learned to develop an intuitive understanding of basic components. An important part of this was developing the skill of estimation, to get an idea if the circuit is even close to where it is supposed to be.

Estimation plays an important role in troubleshooting too. If you are good at estimation, your intuition will be correct and will point you down the right path to solving the problem. Combine that skill with the power of the modern day calculator, and even a circuit simulator as we talked about in Chapter 5, and you have a powerful toolset to diagnose the root cause of the problem.

Can You Break It Again?

This is a simple rule that is often overlooked. Once you have found and corrected the problem, can you break it again?

That is, can you remove the fix and see the circuit wig out again, doing whatever it did before? Often, especially with problems that are difficult to repeat, an engineer will apply a fix, have the problem seemingly go away, and figure he is good to go. However, if the problem is a bit temperamental, meaning it doesn't always show up when you want it to, you might just coincidentally apply the fix when it went away on its own. It my experience this can happen quite often, so I recommend breaking it again to see if you are really fixing it or not.

It's no fun to think you have fixed a problem, only to fire up the production line and shut it down again when the problem reappears. You can also spend a lot of money applying fixes that are not really needed. So please break it and fix it several times before you are sure you have it solved.

Root Cause

A good troubleshooter will methodically trace an offending signal back to its source. As he does so, he will question each component in the circuit as to whether it is operating correctly. He will ask himself things like "Does the output signal of this op-amp agree with the signals that are on the input pins?"

This is why the really good engineers seem to always be muttering to themselves. They aren't schizophrenic, they just ask themselves a lot of questions. (OK, maybe they are schizo, but trust me, it's in a good way.)

Eventually you will find the root cause of the problem—the component that isn't doing what it is supposed to—and then you can figure out why and get it corrected.

Categorize the Problem

Good troubleshooters will separate the problems into various buckets and use an approach that works best for the type of problem suspected.

> *Design problem*: This is the most common mistake and the easiest to find, as it is generally repeatable and consistent.
>
> > Approach: Since it is repeatable, keep it misbehaving while you use tools (scopes, meters, etc.) to trace down the problem. Make sure you get to the root cause.
>
> *Tolerance problem*: Really a design problem, but I give it a special category because this is typically inconsistent and difficult to repeat. Environmental effects commonly aggravate this type of problem.
>
> > Approach: You will need to repeat the environment that caused it if possible. Here is also a good place to run simulations where you can vary the tolerance of the parts you suspect and see what happens.

EMI problem: This can also be difficult to repeat. Who knows when EMI is going to hit? It will often trip up the most competent engineers.

Approach: This one is so much fun I have dedicated a whole discourse to it, coming up next!

Software problem: So many products today use some type of software or firmware. I have seen software exhibit all of the symptoms above and also be used to correct some of these problems, even though it was really a hardware issue. It gets its own category for that reason.[2]

Approach: Give up, go home. Not really, but it is a fact that these can be a bugger (pun intended) to figure out in a reasonable amount of time. Combine that with the natural fear of oscilloscopes that software engineers seem to have[3] and you can see you are in for a treat when diagnosing a software problem. The longer this paragraph gets, the more I think it needs its own discussion, so I put one in.

Go Shoot Some Trouble

Now that you have some basic skills, put them to the test—take aim and blow that trouble out of the water! As one last idea, keep notes of what you are looking into and the conclusions you are drawing. This is especially important if what you are looking for is taking a while to find. It is also nice to have when you are creating your design guides. You can refer to these notes to know what not to do in the next design.

I know it sounds like those dastardly lab books you had to keep in school, and it is, but remember you aren't getting graded on them. Just keep them in a way that make sense to you. Take some notes, get out there, and blast trouble away.

---

[2] Here is a metaphorical question that will drive your code junkies nuts. If you can fix a hardware problem with software, was it really a software problem in the first place?

[3] Maybe I am wrong, but it seems like I am constantly reminding the SW engineers to get out a scope and have a look at the signals they are making.

Thumb Rules

- Do not discount a theory outright; try to prove it right or wrong by experiment.
- Use the shotgun wedding approach to get to the root of the problem quickly.
- Start by checking the simple things first.
- Look outside your specialty.
- Don't ignore anything, and the corollary, don't assume anything.
- Look for what doesn't belong.
- Use estimation and intuition to lead you in the right direction.
- Dig for the root cause.
- Can you break it again?
- Categorize the problem and customize your approach.

Ghost in the Machine—EMI

Have you ever had a circuit or design do something you don't want it to and you just can't explain why it does it? Worse yet, it doesn't do it all the time, just when the planets are properly aligned? You may just have a circuit haunted by EMI's (pronounced Emmy's) ghost. Dealing with EMI is definitely a school of hard knocks course. Here are a few "Cliff Notes®" for those of you that are about to enroll.

EMI stands for *electromagnetic interference* and, boy, does it ever interfere! I remember one of my first bouts with this ghoul. We had recently completed a design of a display that worked great on the bench and even worked most of the time on the product. However, about 20% of the time when you turned the motor on, the display would simply freak out. By an all-night process of trial and error, we finally stumbled across a solution to get production up and running again. Since then, I have learned a lot about how to pinpoint an EMI problem and resolve it. The things I point out here work well when combined with the troubleshooting techniques previously discussed. Few engineers have ever dealt with EMI on anything other than a troubleshooting basis. Let's face it, we don't go looking for EMI, it does just fine finding us by itself! Let's start by getting a basic understanding of what EMI is.

What is EMI?

EMI is basically an unwanted signal entering into your circuit. It is still an electrical signal, it still obeys Ohm's Law and, for all its supernatural behavior, it is still just a signal. This is good news! It means you can exorcise these demons from your design because they still obey the laws of physics.

The Ways of the Ghost

First of all, how does EMI get into a circuit? There are only two ways: conducted and radiated. In the first case, the unwanted signal has to travel on a trace-wire or other path into the area of disruption. In the second case, the signal propagates without wires. It is important to know how the signal is getting in because that affects the solution you will need to employ.

Conducted EMI

How do you know if it is conducted EMI? The easiest thing to do is disconnect everything part by part till it goes away. Case in point, we were hooking a computer up to a circuit board, both at the audio output of the sound card as well as the serial port. There was an annoying buzz in the speakers that changed tone in sync with the displays on the board. When I unplugged the serial connection, the buzz went away. We had what's known as a ground loop. This is a specific type of conducted EMI. I usually try to detect whether the problem is conducted EMI first, as this is the easiest to check. Don't overlook the connection to a wall outlet if the AC line powers your device. I once saw a design disrupted every time an overhead projector was plugged in.

Radiated EMI

The best way I have learned to determine radiated effects is to divide them into two camps, the *near-field effects* and the *RF effects*.

Near-field effects can be easily divided further into current and voltage disruptions. Consider this rule of thumb: anything within a wavelength is near field and anything outside that range is RF. Inside the near-field range, magnetic fields induce current fluctuations into a circuit and electric fields produce voltage fluctuations.

Here is a simple test with a piece of equipment that you are likely to have on your bench. Take your oscilloscope probe and, with the ground dangling, move it near an AC outlet. Adjust the voltage range and quickly you will see a nice 60-Hz sine wave. This scope configuration is basically a dipole antenna and it responds well to electric fields.

But what about magnetic fields, you say? Magnetic fields are caused by current flow. By now, hopefully, when you hear current and magnetic field in a word association game, you come up with the answer "loop." So let's turn our scope lead into a loop antenna by clipping the ground to the probe tip. You will see that the previous voltage signal from the outlet disappears. However, move your new sensor near the power cord of the scope you are using or some other device that is moving current.

Voila, you pick up magnetic fields with this configuration. You can often use this simple technique to determine the type of EMI you are dealing with. (And you didn't have to buy expensive sniffers and spectrum analyzers!)

Once you get over a wavelength away, the prominence of one field over the other tends to disappear and you are dealing with RF or radio frequency interference. How do you find out if the problem is RF? Try moving the suspected interference source over a wavelength away and see if you still have a problem.

To sum it up, radiated EMI can be divided into three categories: near-field magnetic, near-field electric, and far field or RF. The only reason to do this, though, is to identify ways to eliminate the problem. In all three cases at some point the radiated effects have to turn into a conducted effect in order to disrupt your circuit. The trick is to stop that from happening.

Deal With It

Whatever the source, at some point in your career you are going to have the opportunity to exorcize the EMI ghost from your circuit. Before we get into specifics, such as when and where to hang a juju bead[4], there are some basic concepts that will help put these demons back in their bottle.

Break It to Prove You Can Fix It

Remember that EMI is caused by some sort of electromagnetic field either conducted or radiated. Often this only occurs on an occasional basis. That in itself can make it hard to track down. So we will review the concept of breaking it. If you ever think you have solved a particular problem, you will need to remove the solution, and see if the problem comes back. Break it, fix it, and break it again, as we learned in the last chapter. Due to the sneaky nature of EMI it is particularly important in this case.

---

[4] "Juju bead" is a term I use to refer to ferrite beads and clamps. It seemed appropriate in reference to the way ferrites seem to magically eliminate an EMI problem.

Here's an example: One time I was trying to eliminate a flickering problem on a display we were using. As I worked out what was going on, I tried putting a ferrite on the wire harness. The problem went away. Thinking I had it solved, I instructed the production line to install ferrites on all the machines. You can probably guess what happened. Shortly after the line started up again, the flicker was back. I later discovered the problem was caused by motor brush arcing, I just happened to put the ferrite on when the motor brushes "burned in" eliminating the noise source. Now I will always remove and reinstall the fix several times to be sure the problem returns and is eliminated consistently. The first thing I ask any engineer when he/she returns with a fix is did you remove it and make sure the problem is still there?

If you can't break it at will, you can't be sure the fix is legit.

Timing is Everything

Another item I have learned is to track down the sick circuit right when it is failing. Often you may be tempted to leave it till you have time to research it. Then when you go looking, you can't find it because it's working now. You have to catch it in the act, so to speak. So when it happens, don't wait, grab your "juju kit" and go ghost hunting. Don't be surprised if something happens on the production line that you can't get to repeat in the lab. Go to the line and try to figure it out. Amazing amounts of noise can be found on the production floor. (Especially if you have welders out there.) Our production line once found a metal table that would mess up a CD player whenever it was within about 2 inches of the table surface. The table was grounded to a steel post holding the ceiling up. I learned you can have upwards of 50V of noise between ground in the outlet and the steel in a building that is tied to that ground. Tying the table to the outlet ground made the problem go away. I didn't forget to try to break it by removing the fix. In fact, I did this several times just to be sure it really was the problem.

It is difficult to get an EMI problem to occur at will, so don't be afraid to go to the problem where and when it happens.

Under Pressure

Sometimes we are under pressure to develop a solution fast. To do that you might try throwing everything you've got at it at once. If you solve it, then try removing one piece at a time. EMI problems are often combinations of various things. If you try one fix at a time, you may overlook a combination of fixes that would have solved your dilemma. You might need that 0.1-µf cap on the AC line and the ferrite clamp on the data harness. As often as not, you will need more than one fix to solve the case.

Be Prepared for Surprises

An across-the-line AC cap will do great things to filter out noise coming into your system. That's why they put them in surge suppressors. That was an absolute truth for me till just a few weeks ago when I was tracking down a noise problem on a communications harness and I noticed something funny. I was observing the noise on the communication lines when I asked one of my engineers to plug the unit under test into a surge suppressor, instead of directly into the wall. The noise got worse. I'm still not sure why, but we used it to improve our filtering and the reliability of the data. Don't make any assumptions. Test everything.

Not All Components are Created Equal

What is X_c for a 1-µf cap and 0.01-µf cap at a frequency of 1 MHz? Let's see, $X_c = 1/(2 * 3.14 * 10M * C)$, so multiply, cancel the exponents, mumble, mumble, grunt, grunt. You get 0.016Ω and 1.6Ω, respectively. The larger cap should effectively short more noise to ground. Too bad this isn't a perfect world, or that would be the case. Take a look at a regulator data book; what are the recommended capacitors? One large and one small one, right? The reason is the larger capacitors often do not work like smaller caps at higher frequencies. A perfect cap would, but, alas, there are no perfect caps, only perfect calculations. Hint: select a cap with a roll-off close to the frequency you are trying to clamp.

One other thing: the capacitance printed on the case is only legitimate when used at the operating voltage on the case of the cap. The moral of the story—you may have the right component, but the wrong value. Nothing a little experimentation can't solve.

Controlled Environment

Every engineer knows the importance of a controlled environment to determine the validity of a test, yet I see this concern overlooked often when trying to track down an EMI problem. Maybe it is because EMI is so difficult to reproduce. There are some standard techniques for reproducing EMI in a test environment. If you have ever dealt with the European CE requirements, you may be familiar with some of them, such as EN 61000-4-4. This references one test that I find particularly useful, the EFTBN test. It stands for *Extremely Fast Transient Burst Noise*. This is a great test for finding immunity problems with a given design. The history of it goes back to the '60s and '70s. Some IC-based clocks that were being developed seemed to get inaccurate during use. No one ever really located the source of the noise, but they found if the clocks could past this test they developed, they kept time correctly. What they had developed eventually became the EFTBN test. (It creates a similar noise profile to the showering arc test that UL used for some time before replacing it with the EFTBN test.) Remember the rusty file test from Chapter 4? This is the legit version of that.

In the same standard, you can find other test protocols, including static, line surge and others. As you look into these standards, you will find that even the humidity of the room where the test is performed can make a difference. Fully equipping a lab to be able to perform all these tests can be very expensive, but if you do not, don't be surprised by some variation in your results. My own experience with static testing shows it to be one of the most difficult tests to repeat and get the same results. I have seen a circuit tested and seen it pass one level, only to repeat the test on exactly the same board at a later date and get a different result.[5]

One word of caution: just passing all of the immunity tests is no guarantee that your design is good to go. There may still be problems that plague you. In this case you will need to develop your own internal tests that you need to pass to guarantee correct operation.

[5] The moral of that story was circuits will pass static easier on more humid days.

Poor Man's EMI Tests

As we discussed a moment ago, it can be very expensive to set up a completely controlled test lab. Renting time at one isn't cheap either. So what do you do if you don't have much of a budget? Throw your arms up and forget about it? While that is certainly appealing (especially when you are really stumped on a particular problem) it usually isn't an option.

There is a rule that crops up time and time again in every discipline that I have studied. It is the 85/15 rule (you may have heard 80/20 or 90/10). What it means is that it takes as much effort to get 85% of what you need as it does to get the last 15%. This is true in the world of EMI as well. Even if you do not have a perfectly controlled environment, you can still learn something about EMI. What you will not get is a definite pass or fail conclusion.

I have already mentioned the rusty file test as a cheap and dirty version of an EFT machine, but it's not as controlled or even anywhere close to being as safe. It is a poor man's version of the showering arc test. (The showering arc test was used by UL for some time before it was replaced by the EFT test.) I take no responsibility for injury caused by being so poor that you have to use the rusty file test, and I do not recommend it. Personally I think you should get your company to cough up the money for an EFT machine. You will have to spend a few grand but you can get a lot from that without all the expensive shielding room and environmental control equipment. Besides I will sleep better at night if I don't have to worry about engineers rubbing wires on rusty files.

I have heard of cheap and dirty static tests using piezo igniters out of barbecue grills, they pump out 15 to 20 kV in a static jolt. You can get about 5 to 10 kV with a nice pair of lycra shorts on a dry day. (Beware though—you might get some funny looks from coworkers if they see you shuffling around in your biker shorts and stocking feet carrying a PCB to test.)

Again, you can purchase a static gun for a lot less than you can get the whole humidity-controlled room with grounded floor, and get 80% of the controllability that you need.

Line surges can be created by switching AC motors on and off with a simple switch. An AC fan from Wal-Mart is a common source of this type of noise. Again you won't be able to control the level, but you will get an idea of whether or not your design can handle EMI at all.

In general, you should do what you can to check your design. If possible, spend some money for some equipment to test, but you don't have to dive in whole hog to get some benefit out of EMI testing. This way, you can do most of the improvements at your lab, saving time and money when you take it to a certified testing lab.

I Dream of Juju

Experience is of great value in the battle against EMI, but you don't have to learn all the courses the hard way. You can learn from other's mistakes. Read what you can on the subject, but beware there are many different opinions on this topic. Don't take what you find as gospel in your particular situation. By its nature and complexity, EMI can be a bear to handle. You will find some solutions won't work as well for you as they do for other people you read about. The best way to deal with this is to document your reasons and conclusions for a given fix you have found, refer to it and update it often. Make yourself a "Juju journal." (Yeah, sounds a lot like keeping a lab book, doesn't it?) You will find after a while that there are some solutions that work particularly well for your product. Armed with this information, you will solve these problems faster and cheaper than before. You will even begin anticipating avoiding them after a while. Yes, I have even woken up in the middle of the night with the solution in mind. Don't overdo it though; you don't want all your dreams to be of Juju beads and PCBs.

It's In the Air

If you are trying to stop EMI out in the air, your most likely solution will involve some type of shielding, which means putting your design in a conductive box. If it is RF, you will need to keep the holes in the box smaller than the wavelength of the signal you don't want.

If it is near field, there are some variations on the box. Sometimes all you need is a grounded plate between the circuit you are trying to protect and the source of the noise. For magnetic fields or current effects, ferrous shielding works well. For voltage or capacitive effects, something simply conductive will work. Whatever your approach, if you try to stop it in the air, it will involve some type of shielding and very much be a trial-and-error process. It is also the most costly solution. For this reason, I tend to treat shielding as a last resort. I go to the wire first.

It's In the Wire!

At the end of the day, all EMI is conducted. EMI can't disrupt anything until it is conducted. Even when you are dealing with near field and RF disturbances, when it is all said and done, unless it disrupts a signal on your board, it doesn't matter. That alone makes learning how to deal with conducted EMI important. It also means that the board and circuit design itself can affect EMI tremendously.

There are some rules of thumb in PCB and circuit design you can use to stop EMI in the wire.

Low Current Signals are Disrupted Easily

Signal-to-noise ratio is based on power, both voltage and current. Mostly we work in a world where we keep voltage the same and current is allowed to vary. That combined with a need to conserve power often leads to some very low-current signals. The problem is, if the signal is low in power, the corollary is that it won't take much power to disrupt it.

For example you can stick your hand in a stream from a 49-cent squirt gun and easily deflect the water, disrupting the signal. Try doing that with a fire hose and you might lose your hand.

In most cases radiated signals don't have much power behind them once they are absorbed into your circuit. That makes it easy to combat them in one simple way. Make the circuit under distress use more current and thus more power—turn it into the fire hose so it can't be easily disrupted.

Take a sensor with a 1-meg pull-up at the end of a 4-foot wire. Change the pull-up to 10K and watch what happens. This is one reason the old 4/10 mA current loops are so darn robust. They are hard to disrupt.

If you really can't spare the extra current, you will need a component that has a low impedance at the frequency you are trying to suppress and a high impedance at the lower frequency your signal is operating at. They have those; they are called capacitors. Putting one of these on back at the input of the device in question will create a load at a specific frequency, making it harder for the unwanted signal to disrupt the wanted signal.

Find the Antenna and Break It

Increasing power to a circuit works great unless the signal causing you fits is at the same frequency as the signal you need to read. When this is the case, you need to consider antennas.

In a very real sense in the world of electronics, everything is an antenna. The only question is, how good of an antenna is it? But first what *is* an antenna?

An antenna is a device that turns a radiated field into a conducted signal. There are two basic types: the dipole, a ground and a length of wire, and the loop—you guessed it, a loop of wire. Earlier we learned how to turn a scope lead into both types of antennas to discover some of the EMI in the world. The loop is particularly good at picking up magnetic effects while the dipole does well with capacitive effects. At RF levels, there are all sorts of equations and loading formulas that are more in depth than the scope of this text. Suffice it to say that RF can be picked up with both antenna types.

The trick is identifying antennas in your design. Once you find them, then you can figure out what to do with them.

Sometimes you might identify an unknown antenna in your circuit when you are checking for conducted effects. You might unplug a long wire, for example, and discover the problem goes away. I have had this exact thing happen more than once where I unhooked some contacts that were getting a static discharge, only to still have a problem. The problem only went away when the wires that routed out to these contacts were unplugged. I had removed the antenna.

Dipole antennas tend to be wire harnesses that plug into the design. One way to hamper these antennas at higher frequencies is to put a ferrite bead on them. Now you know why those little bumps are on so many wires these days.

Loop antennas are often found right on the PCB. The higher the frequency, the smaller the loop needed to have a problem. In general, the smaller these loops, the better your design. An easy way to improve this, if you have money to spend, is to go to a 4-layer PCB with a ground plane and Vcc plane on the center two layers. That way you always have the smallest loop area. If you don't have the bucks to spend on a 4-layer board, it will take some practice and patience to learn how to do the same thing with a single or double layer PCB. I highly recommend a class on this topic for your PCB designers if this is the case. There are many available lecturers on the subject.

As a general rule, good radiators are good receivers. This being said, you can turn your circuit on and using the scope probes, find hot spots on your PCB or wire harnesses and get an idea of where the trouble is. If you need to be more precise, you may want to invest in some near-field and sniffer probes for your equipment. Find the antennas in your circuit and break them (make them bad antennas) to stop EMI.

In Conclusion

There is no simple approach to dealing with EMI, and experience rules in this arena, so don't be afraid to get your hands dirty trying to figure this out. Also there are many texts out there on this topic and this is by no means comprehensive, but I will warn you not everyone agrees on the same approach. You will need to find out what works for you and your product and go with that.

One final note is that the things you do to keep EMI out will also keep it in when you are trying to pass those emissions standards that seem to get tougher and tougher with no end in sight.

Make your circuits more difficult to disrupt, ferret out those unknown antennas and break them, and when all else fails, shield it.

Follow these thumb rules to help you exorcize that ghost in the machine.

Thumb Rules

- EMI comes in two flavors, conducted and radiated.
- Radiated effects can be divided into near field and RF.
- Near-field effects can be magnetic or electric.
- Identifying the type of EMI you are dealing with can help you develop a solution.
- Start with unplugging and unhooking whatever you can.
- Is the fix repeatable? Can you break it by removing the fix?
- Chase down the problem where and when it is happening.
- Remember components aren't perfect.
- Keep a log of solutions.
- Low-current signals are disrupted easily.
- Find the antenna and shut it down!
- Load the dipole antennas.
- Minimize loop area on the PCB.
- Good radiators are good receivers.
- When all else fails shield it.

Code Junkies Beware

Our world relies more and more on software. In saying this, I include firmware, which is really software that you just don't change as often. It is in everything. Even good old analog circuits are evaluated by software in most cases. This is a good thing because of the flexibility that it has created and the new features that are available (my home stereo wouldn't be the same without DSP!), but it comes at a price. The world of buggy software we live in today is that price.

Bug-Free Software May Be Impossible

If we are talking 20 lines of code, we can make it bug free, but what about a million lines? Or even a thousand? The more code there is, the harder it becomes to make it bug free. I have no proof as Einstein did, but I think it is akin to the law of relativity—the closer you get to the speed of light, the harder it is to get there, basically making it impossible. Likewise, the more code you get, the harder it is to make bug free.

Whether your code is 50% bug free or 99% bug free depends primarily on one thing—how much time you have tested it. The more features and complexity in the code, the more time is required. At some point you have to figure out a balance between a level of bugs you can live with and when you need to ship the product. Since we as consumers demand everything now at the lowest possible price, we have created a world of upgradeability. You can buy my possibly buggy program now, and upgrade it later. This even happens in everyday devices, not just computers. I have upgraded my PDA several times, and I just found out there is a new version of OS for my PSP. I have even upgraded my GPS unit a couple of times.

So, if your code is gargantuan and you want really bug-free stuff, your cost will be high and it will take lot of time. Space Shuttle code is up there in the bug-free realm, and it is possibly the most expensive code per line ever written.

This is why big SW companies that start with letters like M sell you code that you never truly own and aren't responsible for it malfunctioning. To guarantee it would simply be so expensive that no one would ever buy it. Software never can be truly perfect, but it can be good enough. "Good enough" is completely objective, however, and it is up to you and your company to determine what level that is. Here are some ways to troubleshoot your code and help you determine if it is good enough to ship.

Testing, Testing and More Testing

Good code takes a lot of testing if you hadn't gotten that idea already. I particularly like human testing, where the person that is using it is involved. We humans always seem to discover ways to break stuff that you just didn't think of when you designed it.

The problem with human testers, though, is getting them to remember what they did when it broke. Memory can be a fickle thing, and when you are drudging through a test, exactly what you did when the unit malfunctioned is likely a poor recording. In one place I worked, we put cameras in the test lab to watch the human testers so we could back up the tape and look at what happened. It saved us from chasing down more than one dead end.

Repeat the Problem

Like most difficult to trace problems, the ones that are hard to repeat are the hardest to find. With software, it is not unusual to have a certain set of conditions required for the bug to manifest. Certain key-press combinations, or maybe timing. If you are chasing down a bug and you just happen to make it repeat, stop, rewind your brain about 30 seconds, and see if you can do it again. Keep trying slight variations on whatever it was that made the bug show up until you get it to happen again, and then try to repeat it one more time. Keep trying till you can get it to happen whenever you like. If you can get it to happen on cue, you will be able to track it down much easier.

Setup Tracers

In code it is possible to set up tracing registers that can keep track of key information that will help you figure out what went wrong. This can take up some extra time in development, but will pay huge dividends in the debugging process.

One time we had a problem with a control panel resetting at apparently random intervals. We checked the stack by creating a register that kept track of how deep the stack would grow. As we watched it, the stack would get so big it would overwrite other areas of the code and it would go into "la la land" till a watchdog timeout reset it.

Often you can use an available display to show this information. However, there are times when you might want the information faster than the display can update, or maybe the display can't show you what you want to look at. In this case you should set up a D/A—some type of circuit or signal that can take any register in your micro and turn it into an analog signal that you can hook a scope up to.

You have to debug this and gain trust in it before you use it. Do so by loading any number into a known register and look at the scope and see if it is what you expect. Once it is working well, you can use it to do the same type of root-cause analysis as the HW guys. You methodically plug each number into it at various stages of calculation and work your way back from the offending output till you find the cause of it all.

This method can be used with simple RC circuits, serial D/A or any myriad of options. Some chips even have some tracers built right into them. The point is to follow the same root-cause analysis as previously discussed, but in this case you have to have an idea what is happening inside the chip at any given point in its processing.

Break It Again

Just as we already learned, this is a great way to make sure you have fixed the problem in software as well as hardware. If you can break it and fix it at will, chances are you have found the bug. This is much easier with the advent of flash chips. In the old days of OTP manufacturing and EPROM prototyping chips, you had to wait forever (nearly 20 minutes, can you believe it?) for them to erase under UV.

Hunting Bugs

Even though I'm a diehard analog guy at heart, last time I looked, software isn't going away anytime soon. So we do have to live with the fact that my DVD takes longer to boot up and read a disk than my TV took to warm up its tubes 30 years ago.

The fact is code has become a way of life. We are even teaching our children how to handle the convoluted and twisted thinking you need to write code. Just take a look at the video games they are playing! I think I need an upgrade to my noodle just to play them, and I dumped more quarters than most of my peers years ago.

Enough reminiscing. Software is here to stay, and unless the Internet gains consciousness sometime soon and can debug itself, it is up to us, so good luck on the bug hunt!

Thumb Rules

- Test a lot, record info somehow, don't rely on human memory.
- Rewind your brain 30 seconds and try to repeat the problem.
- Set up tracers, use what the chip has, if not, build in your own.
- Repetitively break it and fix it to prove you have found the bug.

CHAPTER 7

Touchy-Feely Stuff

This is the touchy-feely part of the book. Before you say "Ick!" and chuck it as far away as you can, please read on. Most "average" people find those that gravitate to the world of electrical engineering a strange lot. If it weren't true, Dilbert just wouldn't be funny. From the point of view of the EE, the rest of the world often just doesn't seem to "get it." If you want to be the most successful engineer you can, there is some touchy-feely stuff you ought to chalk up on your list of acquired skills. Yes, it is extremely likely these are going to be acquired skills; the engineer that comes by these capabilities naturally is a rare breed.

People Skills

One difficulty engineers often have in dealing with people is the fact that interactions between us human beings can't be described by slick mathematical formulae like the various circuits they are working with. I personally think this is why you often see engineering groups managed by nonengineering types. So what should you do? One thing I have found is that, while there is no perfect equation to describe people, there are some categories that you can sort people into to help you understand how to interact with them.

In any business organization there are levels of hierarchy—there is no round table. Some one sits at the head and it goes down from there. There is always a pecking order, even if it isn't on the org chart. Let's sort the personality categories into various levels of interaction, as that will definitely affect how you should react. We might as well start at the top.

Note that I am using masculine pronouns in these people descriptions for convenience only. Of course all of these people can be either male or female. Maybe someday we'll invent some effective gender-neutral pronouns. Until then, please feel free to use the pronoun that offends you the least or makes you laugh the most.

Those Over You

The means your boss, the person you report to, and the person that takes responsibility for what you do. Of course that is in a perfect world.[1] First, some general rules:

- Avoid talking smack about your boss. Even if he deserves it, constant griping and complaining will usually hurt you more than him.
- Maintain integrity. Sometimes lying and deception can get you ahead in the short run, but in virtually every case it will come back to haunt you.
- Help your boss succeed. This can be hard sometimes, especially if your boss never gives you credit, but even if that is the case, be a great employee. Someone will notice.

Following are descriptions of some boss types.

The Dilbert Boss

This is the clueless boss. He has no idea what you do, and he is more concerned with his position than the success of the company. He is more than willing to sacrifice one of his employees to make himself look good. This is the type to take credit for everything you do right, and blame you for everything that goes wrong. First, do the best job you can. His own self-interest will keep you around if you are a valuable employee. Second, look for opportunities where others in management can see your skills. This will counter the fact that he tries to hide you away. Transfer out of this group if you can, as it will be difficult to get far with this boss.

[1] I am well aware that there are plenty of boss types that will take all the credit when you do good, and lay all the blame at your feet when you screw up. I truly hope you are never saddled with such a boss, but read on for some rules that will help if you are.

Negotiator Boss

This is the salesman type, the supreme negotiator. He will always set the goal beyond any reasonable point, figuring that somehow this will encourage you to go further than you think you can. First, don't be discouraged by these requests. After that, you have two approaches you can take. Be a negotiator yourself—overestimate the time and money it will take to get the job done so that you have room to negotiate. (Like Scotty does for Captain Kirk.) The other option is to say what you can do and stick by your guns. Try not to underestimate with the negotiator, though—he will be disappointed when you don't meet the goal you said you would. The negotiator is not necessarily a bad boss to have. You could do much worse. "Better to aim for the sun and miss than aim for a cow pie and hit it," is the creed of this boss.

The "Yes Man" Boss

The yes man is the submissive boss. He tells his boss anything he wants to hear and will often not defend his employees. It is not unusual for this boss to commit you to impossible dead lines and tasks. Don't make the mistake of being a yes man to a yes man though—that is a disastrous combination. Let this type of boss know what it is really going to take. If you have a strong personality, you can help this boss by standing up for him if he does say what it's going to take to get the job done to his boss. Generally this boss will give you the credit for both your successes and failures.

The Micro Manager

This boss tries to manage every detail. Try to handle his status report requests and required updates as quickly as possible so you can get back to business. He may even be so obstinate as to be upset when you try to make a decision for yourself.

I think the best way to deal with this type is to simply make sure you get those reports and updates in on time. Try to be so reliable that this boss will gain trust in you. Often you can talk to this boss easily (there will certainly be enough meetings with this guy). Talk to him about your priorities often, and stay in sync with his goals for the department. As long as they don't carry it to extremes, there are worst things than being micro-managed.

The Macro Manager

The opposite of the micro manager, the macro manager is the boss that is never there when you really need some help. He is hard to get hold of, and often difficult to talk to. This leaves you making a lot of decisions that you may not feel comfortable with. You may even be criticized for decisions you have made after you asked repetitively for some feedback on that particular issue without response.

The best thing to do in this situation is to take advantage of the opportunity to learn to make decisions on your own. You may screw up, but that is a risk you take in any decision situation, so don't be too afraid of making a mistake. If your boss does question your reasoning, try to explain your decision process. Remember he wasn't there for all of the things that lead to your choice. Don't assume he has the background on the issue that you do. The best thing about this boss is the opportunity you will have to shine. You will be given plenty of rope; try not to hang yourself.

The Perfect Boss

The best boss gives you some credit, while buffering you against mistakes, giving you a chance to learn and grow. If you have this type of boss, do your best to succeed and you will! You should hang on for the ride! Often he or she will give you plenty of leeway to succeed. They will recognize that their success depends on yours and they will help you to succeed. Don't be upset if they get some credit for something you did. If they are a good boss, they created the environment that allowed you to be a success. Often as this boss succeeds, you will too because he will bring you along with him.

Your Boss' Boss

You may not get a lot of interaction with this guy, but take care when you do. This is the most visible you will get as an employee. Try not to be too nervous. I remember one time I was dealing with the CEO of the company. Our production line was shut down because of an electronic power problem. I was a lowly part-time student tech in the QC department. I had just figured the problem out when he came to the line to see what was up. I was shaking in my shoes as I showed him the cause of failure. He didn't believe me at first, so I showed him a broken one, fixed it, then broke it again. He was satisfied, and production started back up. It only took two or three more of those situations and the CEO knew my name. If I had panicked in that position, no matter how right I had been, the results for me would have been a lot worse.

Summary

A point to consider with these categories is that it is possible to find variations of these types. After all, as we said originally, this people stuff isn't an exact science. If your boss is a blend of these types, you will probably have to blend your response as well.

Those at Your Level

Ah…your coworkers, your fellow peons and sometimes enemies. This level of interaction with your network of peers is the best place to create future opportunities. Here are some peer types.

The Sneak

Watch out for the sneak. He is always trying to see what he can get away with. He will only work hard when the boss is watching. Don't get caught in any of his schemes to take advantage of the company. That usually turns out bad and gets you branded as a sneak as well.

The Power Monger

A true political figure at work, for him it is very important to build power and reputation. What is sad is that he may try to make you look bad to do so. Try not to give him any ammunition that he can use to prove how bad you are doing, thus making himself look better. You can make alliances with this guy pretty reliably, but it will be an "I scratch your back you scratch mine" type of relationship. If you make deals with this guy, you will need to hold up your end of the bargain, as you will be relying on his own self-interest to hold up his end.

The Badger

This person will tend to respond emotionally to a situation. If he feels he is being attacked he will likely get defensive and angry like a badger when cornered. The best thing to do is back down and give him a chance to calm down. If you can help them get past the emotion (or just wait it out) you can usually reason with them. It is not unusual that the badger is also a workaholic. Maybe that is why they are so ornery…

The Average Joe (or Jane)

Companies are filled with average Joes. These people do a decent job, nothing stellar, but are fairly reliable. I believe if it weren't for average Joes, companies could never be formed and kept together. These guys like the security of someone else making the tough decisions. They will often ask you what they should do.

Average Joes like to look to a leader. If you can gain their respect, others will notice and it could lead to a promotion.

The Shooting Star

These are the guys (or gals) who get it. They work hard, but don't make themselves into a badger. They are reliable and often correct in their decisions. True shooting stars possess integrity and a desire for the company to succeed. They often get promoted as these skills are recognized. This is a good friend to have in a company, but hopefully after reading this book YOU are the shooting star that everyone else wants as a compatriot!

Often the leader and a true mentor, even if the org chart doesn't show it, you should listen to the star's advice whenever you can.

Conclusions

One of the most important things to have at this level is respect, for yourself and for the others you deal with. You gain respect for yourself by following through with what you say you will. Stick to your word. If you make a mistake, say so, correct it and move on. Give others a chance to build respect with you. This mutual respect is a way to build a network of contacts that is synergistic in nature. Here is where you can help each other out. Do each other favors, and be more successful than you would be on your own.

Those Under You

You may be looking for a chance to lead or have had it forced on you. Either way you ended up with some subordinates that answer to you. This is commonly the hardest adjustment for the true engineer type. As these people below him on the org chart interact, he or she will be baffled at the behaviors and personality traits that come out. Here are a few buckets to sort them into.

The Smart Slacker

This person is usually pretty smart and can get a job done quicker than most others in the group. Because of this they might get some free time when others don't. But they don't go looking for any more work—they goof off or spend the time surfing the net or other such things. Usually they are quick enough on the keyboard to get back to looking busy when you walk by. Keep their plate loaded to the brim. If their slacking becomes a big problem, you may need to call them in and discuss it.

The Praise Deprived

This employee often needs daily feedback on how they are doing. They are looking for positive reinforcement and need a little praise. Be sure to let them know when they are doing a good job. Don't be afraid to be constructive if they make a mistake or should try a different approach. They will usually let you know if they are done with a task and need more to do.

Sometimes as a boss, you will wish they would just leave you alone, as they can seem a little needy. If they are a valuable employee, spend a few minutes with them as needed. They will be very loyal for that little time you spend. If they aren't so good, ignore them and they will find a job elsewhere, solving the problem for you.

The Dud

This is the person that doesn't bring a lot to the table. You have to put more work into him than you are getting out of him. That said, I am a firm believer that people can change and improve. I prefer to give them a chance, but be firm in laying out the expectations. Let him know what is needed from him to keep him employed. However, this is not a situation that you can keep dealing with forever without draining resources from the company. If he doesn't change, this is the person that you have to make a hard choice with, the one you have to let go. Don't run your group with a drain on resources indefinitely. It will hurt all of you in the long run.

The Average Joe

The same guy we talked about above—be a leader for him, show him how to excel and you just might turn him into a shooting star.

The Shooting Star

Same as we already discussed. The more of these that you have in your group, the better you will perform. Don't be afraid of giving them credit, and don't try to suppress them into being your peon. It will backfire on you. Share the credit and hook your wagon to this person and they will get you to the finish line!

Finally

Can a truly effective manager get an average Joe to become a shooting star? Or make a dud into something more? I think so and I believe it is the mark of a good manager to do just that. Anyone can yell and intimidate people into doing what they want. The person that persuades and edifies is much rarer and also more valuable. His team will be more efficient, have less turnover and just get more done. It doesn't mean you should be an old softy. You may need to be firm at times, but if you truly care about your employees, it will show and make a difference.

Administrative Assistants

Every organization has an underground method of communication. In most it flows through the assistants. Building a good rapport with the secretaries and assistants is a good idea. It will allow you to tap into a whole other communication structure. If they think well of you, you will have a better reputation with those above you. Help the assistants whenever you can, and treat them with respect. A lot of unsung greatness is due to the assistants.

Thumb Rules

- Work for the perfect boss when you can, work with what you get when you can't.
- Gain the respect of the average Joes.
- Hook up with the shooting star.
- Be a shooting star yourself.
- Blend your response to blended personality types.
- Give the dud a chance; let him go if he doesn't step up.
- Make the average Joe into a shooting star.
- Treat the administrative assistants with respect.

Becoming an Extroverted Introvert

It seems to me that, generally speaking, the personality types that do well in engineering seem to be naturally shy. I would have to say that electrical engineers are probably the most introverted of the bunch. I was once asked, "How do you tell if you are talking to an extroverted engineer?" The answer: "He is looking at your shoes not his own."

It's funny because it is true. It is also true that the EE can benefit by overcoming this tendency. Here are a few ways to do just that.

It All Depends on Your Point of View

A wise man once said (and I'm paraphrasing) you will find that right or wrong often depends on your point of view.[2] Given that I will try giving you an idea of the way things are seen from the most common sides of the fence. For this discussion we will call the engineer the peon and the manager the pointy hair.

Peon Point of View:

The decisions and directions of management are often as undecipherable to the typical engineering peon as ancient Egyptian hieroglyphics are to the average person. Here are some insights into the thoughts that go through a typical EE's head when dealing with a pointy hair. "Why in the world is this the most important thing now when just yesterday it was the last thing on the list?" Or maybe, "Why can't you understand things like the word 'impossible'?"

In my early years as a peon I coined the phrase, "Management is an unnecessary evil." It accurately summed up my thoughts on the topic. If your manager couldn't help you with fixing that circuit that wouldn't work right, or the code that just didn't execute like it should, what good was he? I mean, sure, he could keep buggin' me all the time about status reports and the like, but couldn't I manage my own time?

---

[2] Obi Wan Kenobi said this; some great life lessons can be learned from Star Wars®!

Even if you find engineers with a manager that they like and think are very helpful, they are still at a loss to understand management decisions. This is often due to a lack of background on the decision process. Good managers will often help this situation with some explanation as to the way they came to the decision. Engineers, while usually a little underdeveloped in the social skills area, still understand numbers and reasons.

There is a natural angst in the role of the engineer vs. the manager. After all, he is the peon in the relationship. The manager at the end of the day is his boss, not the other way around. Remember, engineers spend their whole life asking themselves "Why this?" and "Why that?" It is what they are trained to do; it makes them good engineers. Help answer that question, if you are a manager.

The Pointy Hair Point of View:

First, understand the first and foremost goal of management. It is to make the business successful. It's either that or to make the department he is managing look good, which coincides with the first unless it is sacrificed for the second. (This can happen with bad managers. Hopefully their boss will notice and correct that before it is too late.)

The good manager wants a successful company—how do you do that? It is pretty simple really, you make more money than you spend.

Where the engineer is more focused on accomplishing the task at hand, the pointy hair worries about getting it done on time and on budget. This often puts the peon and the pointy hair on opposite sides of the fence. It is difficult for a manager to understand that unknown things can come up that mess up the estimated schedule the Peon gave them. Actual quote from a manager: "We need to figure out a way to predict unknown problems from happening and avoid them." He was completely serious. To him, that is how to get from point A to point B. To the engineer who is trained to think logically, this phrase will cause his brain to strip a few gears, leaving him generally speechless and unable to respond.

It is not a bad thing to think so far out of the box.[3] If the engineer can shift his head back into gear after such a phrase and look at it as a problem to solve, you will be surprised at what you think up. It is logically true that you can't predict things you don't know. However, you might come up with a way to find out some things you didn't know before, and avoid those. Which is what that "pointy speak" really means.

When two engineers start talking, you will often see pointy hair eyes glaze over as if you were speaking a language they don't understand, which you are. To keep them interested, use words like "schedule" and "budget" a lot. Managers like to speak in absolutes, this will be done in such and such time and cost so much. Engineers like to have some fudge factor. They have seen too many failed lab experiments to believe it will always go right the first time.

In my experience, if you tell the pointy hair it will cost between 10 and 15 bucks, the only price he heard was 10 bucks. This being the case, if you aren't sure you can get to the low price, you'd better not say it, no matter how often he tells you not to sandbag your numbers. If you have some confidence, though, go for it—it is also the mark of a good engineer to get to the committed price and schedule even if it takes some extra effort. Take caution, however—you don't want to sandbag a number so high that you never build anything because it is always too expensive. Remember the goal of a business is to make money, and you can't do that unless you make stuff and sell it.

Talk It Out

If the engineer makes an effort to lay off the acronyms and the manager tries to explain some of the reasons behind his decisions, it will do wonders for your mutual understanding. The most important thing you need is a desire to understand each other. We'll get into the skills a little later.

Visualization

A few years ago, as I watched an interview with Michael Jordan, I realized that we have something in common. No, it is not a 40-inch vertical leap, or the ability to

[3] I've seen pointy hairs so far out of the box that I wasn't sure there was even a box around.

dunk the ball. I realized that for years I had been using a method for success that Miracle Mike also used, a technique called *visualization*.

Everyone who works for a living experiences difficult and stressful situations. It may be dealing with an irate boss, a lazy employee, or a fellow manager that just doesn't seem sane. Have you ever just left a difficult situation in which you were trying to argue your case when you suddenly thought, "I should have said blah, blah, blah or yada, yada, yada?" You might be saying to yourself "hindsight is 20/20," but what I am about to tell you is how to turn that hindsight into foresight.

I remember one of the first conversations I ever had with a CEO. I was a lowly engineering student, while he was the boss of a 700-million dollar company. He hit me with a couple of questions that I was not prepared to answer. I still remember how my mind drew a total blank. Afterward, as I thought about it, I knew exactly what I should have said. I decided that I would not go into such a situation unprepared again. But how do you prepare for something like that? This is what I did.

I started to imagine myself in the situation beforehand. I would imagine how the conversation might go. He would say "this" and I would respond with "that." In my imaginary situation I would try out several different approaches and then imagine a response. I would visualize the person understanding my point and a resolution to the case at hand that I desired. I found that when I did this, the real conversation, when it occurred, followed my imaginary one so closely that I always knew what to say. And better yet, I usually got what I wanted out of it.

You might think I am full of it, but I have used this technique to visualize getting raises and promotions, and I can honestly say that I got what I asked for in nearly every situation. It actually amazes me when I look back at it. I was promoted into engineering positions when I was still a student. I now work with several people, including a former boss, as an equal or superior. I can hardly believe this could happen to a naturally shy person from a hick town in Utah who doesn't like confrontation.

There are no set rules for how to do this, other than the more often you do it, the more successful you will be with the technique. If you imagine it going in 1000 times, the next time you have to shoot that clutch shot, it will go in. Works for Mike, and it works for me. Give it a try.

Affirmations

One of my favorite SNL skits is the one where Stuart Smalley says, "I'm good enough, I'm smart enough and doggonit people like me!" He mocks a technique similar to that of visualizations. It is called affirmation.

If you get into quantum mechanics there is a rule called the Heisenberg uncertainty principle. It was developed to understand some interesting experimental results in which quantum particles (everyday light being one of these particles) act like a wave in one experiment, and like a particle in another. The problem is, they can't be both all the time; the behaviors are mutually exclusive. Anyway, a general conclusion of this principle says that when you measure something at the quantum level, the very act of observation affects the outcome of the measurement. You, the observer, basically get what you are looking for.

Please bear with me for a moment while I digress into very unengineering-like metaphysical rumination. If you get what you look for, can you affect the outcome by looking for what you want? This is what affirmations basically say you can do. Affirmations are a lot like the visualization technique we have previously discussed, but taken to the next level. You not only imagine what you are going to say or do in a given situation, you imagine the outcome you want.

I know it sounds hokey, and I admit that it can be hokey and it is not a perfect process, but I also believe it works. Take any goal you want to achieve, write it down 20 times every day, like "I will get a book published," or "I will get a raise," for example. Give it six months and see what happens. My experience is that it does work; you're reading this book, aren't you? Guess how I started that ball rolling.[4]

One thing that definitely happens when you use affirmations is that your mind spends considerable time pondering what you are looking for. This, I believe, leads to recognizing opportunities when they come your way so that you act on them. Several years ago I had on my long-term goal list a desire to publish a book. It was a goal I affirmed fairly regularly. I thought about it a lot. Then while reading an electronics magazine, I saw an ad for writers in the back. I sent in and they asked for a copy of my work. I

[4] It is no coincidence, in my opinion, that the techniques of visualization and affirmation mirror that of faith and prayer so closely. I think they are principles that simply work.

sat down and wrote my first column. It took and I began writing. One opportunity lead to another and here I am achieving that goal that I had once set out to do. Imagine, however, if I hadn't had this on my mind when I saw that first ad? Would all of this have happened? I don't think so.

You get what you look for, so control your destiny. Say to yourself, "I'm good enough, I'm smart enough, *Insert your desire here*, and doggonit people like me!" Works for Stewart, works for me and it will work for you, too.

Breaking Out of Your Shell

These techniques work very well in helping the naturally shy person to break out of their shell. If you can overcome the natural shyness so common to engineers and learn a bit about the people around you, it will give you career opportunities you might not ever get otherwise.

The hardest part of breaking out of your shell is that first step towards doing it. You have to make the first step. After that each one becomes easier. For example, you need to make a phone call that you really don't want to do. You have already spent plenty of time visualizing it; you have thought through all the possible things that might happen. Now you are stuck—you just don't want to make the call. It is not uncommon to feel apprehension at this point. Don't give up hope though, there is a way through it.

First, clear your head, and stop thinking about what is going to happen and concentrate on one thing, picking up the phone. Once the phone is to your ear, worry only about dialing the number, nothing else. After someone answers, worry only about initiating the conversation. Once it starts, the preparation you did with visualization will kick in and from there on things will get progressively easier.

Repeat

While it will get easier each time you go into a specific situation, these are not skills you can learn once and then forget about. They require repetition, a lot of repetition, not unlike learning to play an instrument or speak a different language. The more you use them, the better you will become at it. Find out the way these work best for you and practice them.

I still encounter situations today where I use these skills that are over 20 years old for me. They still work and I keep finding new ways to apply them. And yes, I still get nervous when it is time to talk to the CEO, but now it goes much more smoothly.

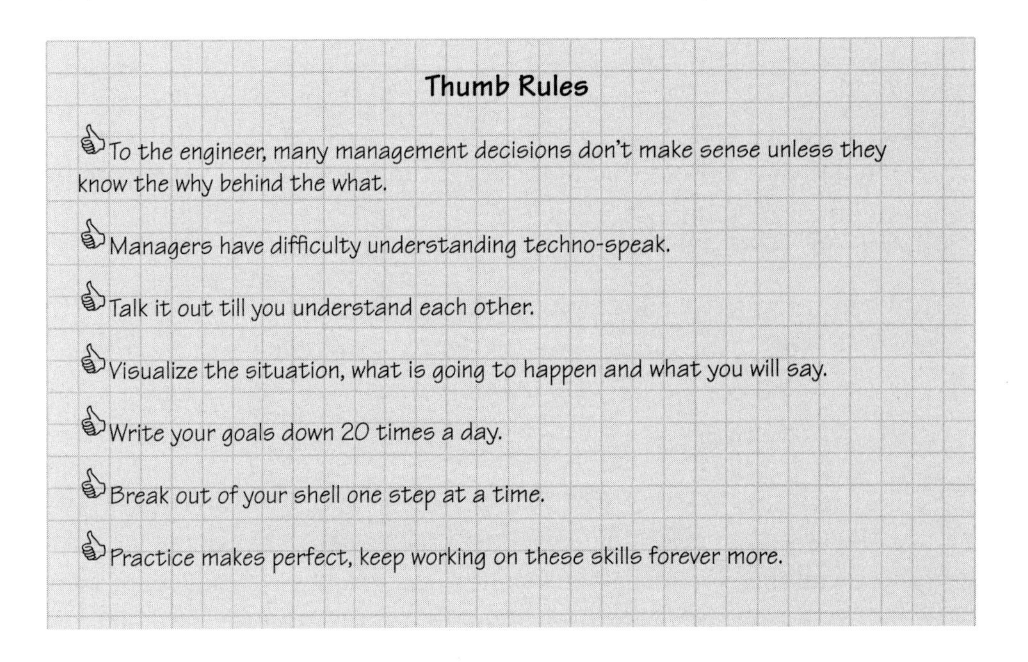

Thumb Rules

- To the engineer, many management decisions don't make sense unless they know the why behind the what.
- Managers have difficulty understanding techno-speak.
- Talk it out till you understand each other.
- Visualize the situation, what is going to happen and what you will say.
- Write your goals down 20 times a day.
- Break out of your shell one step at a time.
- Practice makes perfect, keep working on these skills forever more.

Communication Skills

Engineers are notorious for poor communication skills. I was once asked why it is that engineers that deal in logic that is either true or false have such a hard time answering yes or no to a simple question.

It is something that I myself am plagued with. Given a typical question, for example, "Will such and such project take long?" my answer usually begins with, "It depends..." If I'm not careful from that point on, it can quickly lead to the glazed-over eyes of the one listening to me.

When you are a better communicator, you will be more successful. Simply put, everything we do in the world today requires communication. It is somewhat ironic that things that enable communication to be better (like the Internet for example) are designed by engineers that could often use a class on the subject themselves. So here are some pointers.

Verbal

The brunt of day-to-day communication is verbal. It is also the hardest one for engineering types to handle well. (I think it goes back to that shy thing we were talking about earlier.) However, it is also the most important communication skill to have. Face-to-face verbal communication is the best situation in which you can a) make sure you are understood, and b) make sure you understand.

Watch for Body Language

Some say as much as 90% of what we communicate in a verbal conversation is in fact not verbal at all. There are whole books on this topic, if you really want to get deep into it that tell you the meaning of things like glancing right or left, up or down and all sorts of looks. Most of the time, however, I believe if you simply pay attention you can get a lot out of how a person presents himself and the way he acts. You have been doing this from a very young age and it comes naturally if you give it a chance. Too often we get so rushed or distracted that we don't notice simple signs. For example, a person looks uncomfortable when they agree to a deadline. If you don't notice and dig deeper you could have a nasty surprise coming later.

Consider Who You Are Talking To

The background of the person you are talking to should be considered as you communicate. Don't get caught in the trap of trying to explain details of quantum theory to the CEO who has an MBA. You should try to distill what you are communicating to the points that matter to the person you are talking to. Take note of one important point though—don't ever treat the person like they are dumb! You can distill information without talking down to a person. If invited to you can elaborate. You might be surprised at what your boss can understand, especially if they have read this book!

If you are dealing with a person from a different culture or who speaks a different first language, it helps to simplify your phrases to be sure you are understood. Don't get into vocabulary words that they may not know without being sure they understand what you are referring to. This particularly applies to words that have meaning only in your corporate culture. Everyone perceives the world through perceptions they have based on the culture they come from. You don't need to speak LOUDER. It doesn't help. Try to enunciate your words though, if you are like me you are probably carrying some hick accent that would cause you communication problems in your own native tongue.

Should You Get Angry?

Sometimes getting angry is a correct response. There are occasions when that is what it takes for the person or persons you are talking to to understand the seriousness of the point you are trying to communicate. You may have no other resort to get the point across. However, it should be rare, and if it is rare, it will carry a lot more weight than if you are someone who pops a cork every time something goes wrong.

Reflective Listening

A great way to improve verbal communication is to use a technique called reflective listening. The idea behind this type of communication is to further your understanding of what is being said by repeating it back to the person you are talking to. Take care, however, not to be annoying. No one likes a copy cat. The trick is to rephrase it in terms you understand and see if the other party agrees with you. This is particularly useful in dealing with persons from a different culture, say…a guy from engineering talking to a guy from management, for example.

Read

I believe the single best way to improve your verbal skills is by reading. When you read you experience how others communicate. It works with spy novels to white papers—the more you read, the better you will communicate with others. You will add to your vocabulary, you will understand cultural differences, and you will be able to order your thoughts in a way that is easier for others to understand.

Written

Whether it be emails, reports, or very official-looking documents, writing skills are extremely important in the field of engineering. Considering nearly every engineer I have ever dealt with has had some issue or another with writing, I figure this is an often-overlooked skill.

Proofread It

First of all, writing has one distinct advantage over verbal communication. You can look it over before you print it, send it, post it or publish it. You should proof every written communication you create. The only question is how much. If it is short and you are going to follow up verbally, a quick glance will be enough. On the contrary, if it is going to be read by a superior, or someone who may have reason to pick it apart, go over it a few times.

The most basic skill that I think should be used to proof writing is to read it out loud to yourself to see how it sounds. Don't forget to pause at commas and stop at periods, like you were told to in grade school! This technique will help you root out all sorts of odd-sounding phrases.

If you are particularly concerned, try it out on someone else and see if they understand it. Make sure the person has a similar background to the audience the document is intended for.

Use Appropriate Emphasis

In written communication you lose the ability to create inflection with your voice and you can't tie body language to what you are saying. This can be compensated for by emphasizing what you are saying with fonts, capitals, italics, bold, bullet points and numerous other options available today. If you SAY SOMETHING IN CAPS you create the idea of yelling or raising your voice. **Bold words** can create importance and italics can help you draw attention to *something in particular*.

There is, of course, a whole world of winks, smiles and other punctuation communication out there, but I believe, while most will get the smile, the rest is a code that is known to only a few.

Note that I said *appropriate* emphasis. It is easy to get carried away. Don't cause death by bullets. Too many bullet points and they no longer have meaning. Too much bold and it does no good, too many caps and people will think you are always mad. The trick is to be skillful in applying these skills.

Use Verbal Skills In Writing

Some of the verbal skills above are a great way to improve your writing skills. Things like considering your audience and reflective listening (or reading/writing in this case) can help you understand and get your point across.

Email Specifics

Watch out for flame mail. In the realm of email, it is very easy to be misunderstood, and people often respond with less tact than they might have in person. If you see a flame war starting, I think the best thing to do is talk to the person *in person*.

Watch the CC list, take care to whom you forward what, and always consider that what you have written can be easily forwarded to an unintended audience.

Get to the Point

Both written and verbal communication have a few things in common. One of them is the importance of getting to the point. Use what you need to create the understanding, but don't over elaborate. If ten words will do, don't use a hundred. Here are some other ways to get to the point.

Use Analogy

One of the most powerful methods of communication is the use of analogy. This works well when trying to explain a problem, concept or theory. Analogy helps the person receiving information to visualize what is being talked about. For example, analogy can help a person understand the details of a topic the same way that a telescope can help you see details of the moon. (Or maybe the apartment next door.)

There you go. I just used an analogy to explain analogy, and possibly a little humor too. It is the art of comparing the new idea to something already known. It works very well.

Sketch a Picture

You've all heard the phrase a picture is worth a thousand words. Engineers typically get this; after all, they use schematics which are simply pictures to represent ideas. In the world of email, however, we often ignore what we know so well. We will spend three paragraphs trying to explain what we want when a simple sketch will get the point across. Get yourself a scanner and use it to send a sketch with that email!

Watch Out for Corporate Culture Code Words

Every conglomeration of people will develop code words to speed their communication. In a corporation everyday words will take on completely different meanings. When you are dealing with persons outside of the company, be sure you don't assume they know what you're talking about if you use a corporate word or phrase.

Thumb Rules

- Watch body language.
- Consider who you are talking to.
- Anger is sometimes appropriate, but should be rare.
- Listen reflectively.
- Read.
- Proofread it.
- Use appropriate emphasis.
- Use analogy.
- Sketch a picture.
- Explain corporate culture code words to those not of your culture.

Especially For Managers

Early in my career, I developed an outlook on management that can be summed up in a single phrase I wrote in my day planner in a boring meeting many years ago, "Management is an unnecessary evil." Years later I got my own "pointy hairs"[5] and I discovered a few reasons for management to be. (They may be good and true reasons or simply an attempt to justify my own existence; you will have to decide which.)

The Facilitator

To facilitate is to make easy or easier. Management should be a facilitator; it is up to you to create the environment in which an engineer can succeed. You need to get your engineers the tools they need to succeed. You need to help translate to your superiors the techno garble that engineers are so fond of. (That or buy them a copy of this book and see if that helps.) Most engineers just like to design stuff and really don't want to be in charge and manage things. They like to leave that up to you.

The Buffer

The best managers are a buffer between the top-level antics of illogical requests with unreasonable timelines and the real world of actual schedules and needs. They bring some order to the world of the engineer out of the chaos of pointy hair decisions. This is something the engineer needs to be successful. Don't forget that, in their world, it helps considerably if things make sense.

The Advocate

Good managers understand their employees and are their advocates. If an employee always has to beg for a raise, he will soon be looking elsewhere. If he or she is a shooting star or even an average Joe, you will find a reasonable show of appreciation raisewise is much cheaper than hiring and training a new guy. It is not only right to be the advocate for your employees; it is good for the self-interest of the company as well. I get sick of hearing managers over-talk recognition and promotion as a way to make an employee happy in lieu of a raise. It is true that these things are nice, but

[5] Another reference to Dilbert. It's not a cartoon, it's a documentary.

that only matters if basic needs are being met, needs like food and shelter. If you are underpaying too much, no amount of awards will keep employees around.

Understanding Engineers

Not really a reason for management to be, but knowledge to help you be a better manager. Here are a couple of things you may or may not know about engineers.

Weasel Room

Engineers need a little weasel room. Have you ever asked an engineer if he is 100% confident he has the solution? If you have, you were likely treated to a look of complete loss. I'll bet you were. It is not possible for an engineer to be 100% confident in anything. In this discipline you are constantly assaulted with the fact that you don't and can't know everything. You discover new ways for things to go wrong daily and are constantly working to fix and prevent them from happening. If he gives you a range, take the conservative number for your estimate. Give the guy a little weasel room. Try to pin him down too hard and it could backfire on you.

The Eternal Optimist

I haven't met an engineer yet who didn't underestimate how long it takes to do something. This is simply a fact—good engineers by nature are optimistic, and the really great engineers will push themselves so hard that they will meet the optimistic schedule they set for themselves. I heard a rule of thumb once about writing software that I have found to be true: take the engineering estimate of time it will take and multiply by three.

Desire to Grow

The better you understand the "sparky" viewpoint, the more successful you will be at managing him. If you take this to the next level you can help your engineers take on more and more, literally turning an average Joe into a shooting star, or even possibly rescuing someone from dudsville.[6] Most engineers want to grow and become better

[6] If you don't get the references to shooting stars, average Joes and duds, you either skipped a few pages or have a serious memory problem.

at what they do, but they need a little encouragement, a chance and maybe a bit of a buffer against failure.

The Best Manager is Right Most of the Time

Sometime after I decided that management is an unnecessary evil and then later found some purpose in life after being inducted into management, I came up with a formula that describes a good manager. Remember a manager spends nearly all of his time making decisions, what tools to buy, what people to hire, what to do about a particular problem, what to eat for lunch, etc. How good he is depends on how often he is right. If he is right more often than not, the company makes money. If he is wrong too much, down the tubes it goes. So without further ado…

A good manager is right 51% of the time, a great manager is right 70%, 80%, even 90% of the time. If your decisions are right most of the time you will succeed.

Remember this when faced with a decision. You don't have to be right all of the time. Don't let indecision and too much worry prevent you from making a choice. Often that in itself can cause you to fail. Consider the situation, take action and watch the results. Don't be afraid to admit you were wrong if you see it was a mistake. Learn from the mistakes and don't do them again.

Finding the Shooting Star in a Forest of Average Joes

One of the most challenging things a manager has to do is, after spending an hour or so with a person, is to decide if they would be a good employee and hire them. As we learned in previous discussions, you really want to stock your group with shooting stars, but how do you find them? How do you weed out the duds so you aren't saddled with a problem down the road? While there is no perfect solution, here are some key points to look for in a perspective engineer.

Dress

Don't put a lot of value on how a person is dressed. Casual attire is the norm where I work, so unless someone comes in with serious hygiene problems, I don't chalk up any negative points. Once, however, a potential employee asked what the dress code was. His consideration impressed me. However, it is of minor importance. Our company is interested in results and product, neither of which is significantly affected by the dress of R&D employees.[7]

Fundamental Knowledge

This is very important to me as a manager. There are some skills I don't want to have to teach you, skills I expect you to know for this position. While a degree or some type of schooling tips the scale favorably, I do not consider it a shoe-in. I have seen too many college graduates who got through school by the "assimilate and regurgitate" method. They passed all their tests with great grades, but didn't focus on retaining the knowledge. I weed these people out with questions such as the following:

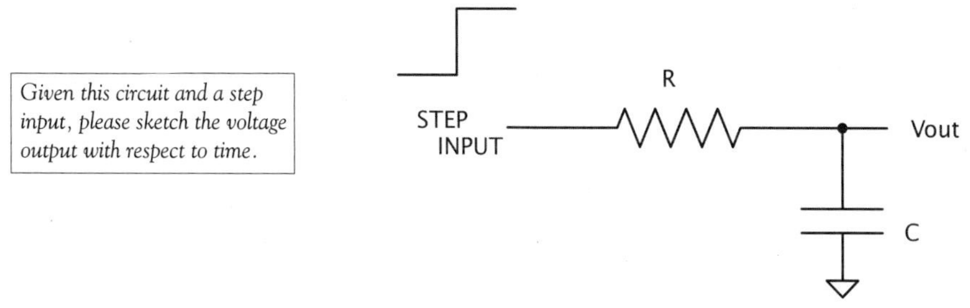

Figure 7-1 *Standard "sparky" interview test question*

You may laugh, but being right next to a major university with a reputation as a good engineering school, I constantly interview fresh graduates who should know this stuff. Fully half of the applicants I see get this wrong! The basics are important. If you don't have them, you are just guessing when you design. Worse yet is if you think you know them and you really don't. After all the hammering on basics at the beginning of this book, I hope the importance of this is understood. I'd rather hire someone with the basics down pat and a 2.9 grade average than the guy that has a 4.0 and stumbles on basic understanding.

[7] OK, that's not entirely true. When I think about it, the casual atmosphere we maintain makes us more productive, but that comes after the hiring decision, so it doesn't count.

Can You Learn?

I have yet to see any employee get into a new job and not need to learn. Sometime during the interview, I will intentionally teach the candidate something new, and then hit the subject later in the interview, to see if he or she has picked it up. This ability to learn quickly and have it stick is important to the success of any engineering group. Technology will quickly outstrip those who can't learn.

Are You Willing to Learn?

You might think this was covered above, but I consider it a separate point. I will often ask the interviewee a question that I am fairly sure they do not know the answer to, simply to see how they respond. Do they try to BS their way through it? Or are they willing to admit they don't know and ask for help? In the rapid design cycles of today, there isn't time to play games. That means, "I don't know, but I will find out" is an appropriate answer. You can take this to a higher level too. On call-back interviews, ask a question again that stumped them the first time to see if they were interested enough to figure out the answer.

People Skills

Is there a job out there that requires zero contact with other human beings? I doubt it. The best engineering teams get along well, which is why I value people skills significantly. How do you handle pressure? Can you get along with persons you don't care for? This is a fairly tough item to evaluate in an interview. I invite my leads to fire questions at the candidate, and watch how he or she responds under pressure.

Attitude/Motivation

A positive attitude always impresses me. I quote my father: "Can't is a sucker too lazy[8] to try." I think it is important to believe something can be done. Look for signs of giving up on a problem. Is the candidate persistent? Does he/she complain during the interview a lot? Do they moan about their last job? I have seen all types. Whiners don't get hired.

---

[8] Can laziness actually be an asset? If it motivates creativity it can. Remember if you give the hardest job to the laziest man, he will find the easiest way to do it. So I guess you could look for motivated lazy people, if that makes sense at all.

Common Sense

This is all about getting the job done in the least amount of time. Too often a person can be book smart, but not be able to apply what he or she has learned. If you don't have common sense you will struggle with applying the knowledge you have. Here is a brain teaser I often use to determine the level of common sense in a person.

> *You are standing in a room with two strings hanging from a high ceiling. If you grab just one string and walk to the other, the second string is several feet out of reach (because it is hanging straight down). Your task is to tie the two strings together. You have just three things to perform this task, a book of matches, two single squares of toilet paper, and a screwdriver. How do you tie the strings together?*

In General

Should you be looking for the person who can do differential equations in his head? I don't think so. I will buy Mathcad® for that. You want to know if the candidate has the fundamentals and if he can and will learn the rest.

Remember, great managers are rare, mediocre managers are commonplace. You don't have to be a great manager for a company to be successful. Why stop there, though? Great managers are huge assets to any company; great managers can turn average Joes into shooting stars can make incredible things happen.

Being a great manager isn't all that hard. Listen, look and learn until you are right most of the time and you won't be unnecessary or evil!

[9] I'm not sure if I want to reveal the answer. What will I use in the future if the book sells well? I will say it has something to do with a pendulum and the screw driver. People that are smart, but with less common sense, will over think it too much. (You wouldn't believe some of the answers I heard!) If you have a high common sense quotient, you will get the solution in a second and wonder why it was so simple.

Thumb Rules

- Management is an unnecessary evil?
- Be a facilitator.
- Be the buffer.
- Be an advocate.
- Understand your engineers.
- Be right most of the time.
- Hire shooting stars.
- Don't be unnecessary.
- Don't be evil.

Especially for Employees

As an employee your motivation, like the pointy hair boss you work for, eventually boils down to money. You want to do a job and get paid for it. True that job satisfaction is important, but that comes way second to the need to buy food to eat and have a dry place to sleep. This means an employee needs to know two things—how to get a job and how to keep a job. This chapter is a guideline to those things.

How To Get a Job

It all starts with the interview. Having interviewed more engineers than I care to remember, I have compiled seven definite no-no's[10] extracted from real interviews. Giggle, laugh, and snicker if you will, but please do not try this in your next interview. The persons in these paragraphs are professionals.

Don't be Condescending

Be careful of how you come across to your potential employer. One candidate I interviewed seemed to really disdain coming to us for a job. It was as if he would work for us if he really had to, but he sure wasn't going to like it. The "you don't have anything to teach me" vibe was very strong. Being an engineer that believes the ratio of what we know to what we don't know is extremely small, I have a tough time with that. This is especially disconcerting when some simple circuit diagrams are requested and you get the response, "Everyone knows that," a little hand waving, and then nothing is written down. I immediately think you don't actually know it, and this is all an act to cover up the lack of knowledge.

Don't Worry About Saying "I Don't Know"

The stress of an interview may make it the toughest place to say "I don't know" but that is not a bad answer. Especially if you follow up by, "I'll find out though." One of the best impressions I had from a potential employee was when he sent me an email afterward that explained the answer to one of our questions in the interview that he didn't know at the time. The fact that he looked it up showed perseverance and a desire to learn. That alone will many times make up for a current lack in knowledge.

[10] Come to think of it, these don'ts aren't just for interviews; you could make a pretty good case for each one as a rule of thumb in almost any job.

Don't Lose Your Cool

One person that I interviewed was clearly thrown a bit off balance by some of the questions asked. What really put marks in the cons column was when he got so upset trying to solve the problem that he threw down his pencil and repeatedly smacked the table. Our work environment can be much more stressful than an interview; I really don't want to worry about someone going mental.

Don't Give Up Easy

If you don't know the answer to a particular problem, try to figure it out if you can. I will often ask questions that I know the candidate won't know, just to see how he/she handles it. Someone who takes one look and walks away has never impressed me. Remember while someone is standing there saying it can't be done, someone else is out there doing it.

Don't Be Afraid to Ask Questions

Along with the point above, you are not expected to know it all. If a person asks a question about a particular task or problem I've given him/her in an interview, it usually shows that a person who doesn't know is willing to find out. That is a very important trait in the engineering world. Also use the interview as a chance to find out about your respective workplace.

Don't Lay Your Head on the Table

Yep, it really happened and I have witnesses to prove it. This potential employee laid his head on the table several times during the interview. I couldn't figure out if he was tired or just listening for some type of table vibration that might indicate how good the interview was going. This would never be my only reason for not hiring someone. (I get some of my best ideas in that twilight between almost asleep and almost awake.) However, this was coupled with some other blatant problems. I just knew it wouldn't work. Let's just say this particular interviewee will have plenty of time to nap now.

Don't call yourself stupid

I wouldn't have believed it if it hadn't happened to me. One applicant we had got a little flustered with a couple of basic questions, but that wasn't what did him in. The first time he said "Man, I am stupid," I didn't think much of it, but as the interview wore on I heard, "Oh, I'm an idiot" and "I am soooo stupid" probably a dozen times or more. By the end of the interview, I was sure of one thing. I definitely didn't want to hire an idiot, especially one so stupid.

A Final Thought

There are a lot of guides out there on getting an interview and getting through an interview. They are even a bit more conventional than my seven don'ts. It can't hurt to study up on some of these pointers. I also think it helps to know a bit about the company you are interviewing with. Take advantage of the ability there is today to look up anything on the internet. It will help you decide where you want to be and it also doesn't hurt to have a little background going into an interview.

How to Keep a Job

When the ax falls, will you be the one to get chopped? How do you increase your stability in a given company? What makes an employer keep one person and let another go? Here are five key areas that can give you a little more security in this layoff-prone world—things you can do besides simply be good at your job.

Value

Here's a thumb rule, *companies exist to make money*. Even nonprofit companies need to bring in money to cover their salaries and expenses. When your employer starts reviewing you and your coworkers, you need to realize that this is foremost in their mind.

This is the question the manager must ask himself. If I had to start all over with just one employee, who would it be? Or in other words, who would most likely make this company a success? In my analysis, this person is the "shooting star." He or she works hard, has great talent, can handle pressure and works well with others. If you ask for something you get it. You don't have to keep checking up on him. You know she is going places. He very directly affects the profitability of the company.

Therefore, you must remember that your total value is of top importance. What if you add value though, and no one notices? This can happen, especially in larger companies. My answer is this. It is not bad to toot your own horn a bit. A good way to do this, both for you and your employer, is to do a regular self-evaluation. List the things you accomplished last year and compare them to what you did this year. Do you show improvement? If not, commit yourself until you do. Then give that to your boss. He'll appreciate that you look at yourself critically and it's a good chance for him to see what you have done for the company.

Position

Repeat the thumb rule we just learned, *companies exist to make money*. They don't do that without a product. So the most important job you can have is one that is directly related to the product. Don't get stuck in a one-off job. What is a one-off job, you ask? A one-off job is one you can eliminate and still sell product. It is one level removed from delivering a product to the customer. The ISO9000 "Corporate Coordinator" might sound like a pretty neat title, but when you get right down to it, the company could do without it. If you find yourself in a one-off job, it's time to start looking for a transfer.

Loyalty

It's human nature to complain. Because of that, an easy yet subtle trap to fall into is right by the water cooler. In this trap you discuss the latest smack about the boss. Every leader I have ever met appreciates loyalty. If you succumb to spreading rumors, whether true or false, you put yourself on shaky ground. I am not saying the pointy hairs don't make mistakes. In fact, I believe that a manager only needs to be right 51% of the time to be successful, as you already know. So remember this, they may have their faults, but so do you. If you have a serious issue with your boss that you can't overlook and can't help talking about, you'd better start looking for a new job, because in today's market, you soon will be.

Effort

This is important for two reasons. First, a great effort can compensate for a lack of skill. Remember that the guy that tinkers in the lab for hours on end can get to the finish line faster than the brilliant engineer who spent the morning surfing the 'net. It's all about getting to the market the fastest these days. It is the entire reason that MAMA exists. All the pointy hairs want to do is to deliver product, make the sale—in general, to do business. So a supreme effort is usually noticed. Remember the same rumor mill you should avoid yourself can have a tremendous effect on you. You can be known for hard work, or you can be known as a slacker. The choice is up to you.

If the Worst Happens

It is possible that no matter what you do, you still get laid off. There are times when a company has to cut deep and there is nothing that can be done. I suggest you take this as best you can, and leave on a good note. If things pick up again, it is a lot easier for a boss to hire someone he knows will do a good job rather than any Joe off the street. So don't burn any bridges.

A Final, Final Thought

By no means do I consider this list comprehensive. There can definitely be more to it. People skills, attitude and other things are considered by an employer when making this tough decision. To make it worse, the world is not all sugar and spice. There are sadistic pointy hairs out there who give the rest of us a bad name (I just hope they are the exception, not the rule). If you have one of those, don't complain, just start looking.

Remember, dealing with people is not a very exact science. There is no Ohm's Law for corporate culture. These are things that I have found generally work. You can sum it all up referring to the different types of employees we have previously discussed, the shooting stars, average Joes, and duds. When it comes time for layoffs, you don't want to be a dud, and if you can help it, try to be a shooting star.

Thumb Rules

- Avoid the 7 interview don'ts.
- Companies exist to make money.
- Companies exist to make money (duplicated to indicate importance!).
- Take care of the 5 key areas.
- It isn't a perfect science.
- Don't be a dud.
- Be a shooting star.

How to Make a Great Product

The slinky, Legos, the PC, silly putty, weed eaters, Velcro, cell phones, DVDs, pet rocks and the microwave—the list of killer products seems endless. How do you go about designing a great product? What makes a product successful? Believe me, the list of great ideas that never went anywhere is much larger that the list of things that made it! For those of you with a more entrepreneurial spirit, here some pointers on how great products come into existence.

The Idea

Usually the core of a great product addresses a need or desire. The more people that share the same need or desire, the more success potential an idea has. Here is a real live example. My car windows are always frosty during the winter here in Logan, Utah. I don't have the patience to start my car early and wait for the defroster to clean the window, so I scrape. Scraping is not much fun, and last year I had a great idea for an invention. Why not put a heater in the windshield washer fluid so I could have a quickly defrosted window without having to scrape? I am sure there are other people like me who would want this product.

Let's evaluate this for a second: first you need to own a car. That limits the primary market to Canada, the U.S., and Europe. Then it has to get cold enough to frost your windows where you live. There goes half the US. Next, to be like me, you can't afford to park your car in a garage and that eliminates a bunch of people. I figure that Canadians like to scrape, knocking off another large part of the market. So will this idea make me a million? Probably not. If I worked hard enough at it, it might generate a decent income for a while though.

Compare that to the market of the weed eater (string trimmer to be more correct). When George C. Ballas stuck some twine in an old tin can and spun it on his electric drill, he was addressing a need that many a man felt. Not only did it chop those pesky weeds, it involved a motor as well and, ohh, the power rush!

His market was anyone that had ever wished for an easier way to trim those hard to reach places in his lawn. To top it off, it also stroked the male ego. I think it had a larger success potential than my defroster idea, don't you? Notice that I said potential. There is a lot more needed to make a product a success.

Design

The product needs to work well. This means the design needs to do what the customer expects from it. If everyone sends the product back it won't be a success for long. There is one all too evident exception to this rule, software! Sometimes people will deal with Glaring Product Faults (also known as GPFs) if that is the only game in town. It is that or you really need a particular feature and are willing to deal with the bugs. It bothers me, though, that you can't send it back because you clicked "I accept" on the 40-page EULA that no one reads, which prohibited you from even taking Bill's name in vain, let alone returning a product. But…here I am using a popular word-processing program because of the features I like.[11]

It also needs to look good. Ever since the 1950s, industrial designers have convinced the consumer that you can have a functional product that looks good as well. There are successful ugly products out there, but if they looked good they would be even more successful. Have you ever said, "That's a sweet little package," in reference to something other than the opposite sex?[12]

Timing

Ahhhh, timing...It is as important in launching a product as it is in telling a joke. I don't think the weed eater would have sold before America moved into the suburbs and the lawn wars began. The slinky wouldn't have made it very far if it came after the Nintendo. A company I worked for had an idea that changed our market place. It was featured on some 30 different news channels and became a raging success. It didn't stick till the third time it was tried. The first two times were utter failures. It needed the Internet as a global community to be a success. The first two times they tried, the global data community just wasn't there to give it the buzz it needed. Timing is important.

---

[11] SW is getting better in terms of stability, but unlike every other consumer product there is still no responsibility for any damage it may cause. All I can say is, whoever thought up the idea of the EULA was extremely smart and darn lucky to make it stick as it has so long.

[12] Possibly a bad analogy, as I have heard an engineer say exactly those words in reference to an IC, but you get the point.

Funding

It takes a million to make a million, right? This is usually the case unless you listen to late night TV. And if you believe that stuff, I have a book on how to get a perfect stranger to give you 50 bucks. I will sell it to you for only 47.95 plus shipping and handling.

I think that funding is one of the things that stop more great products from coming into being than most other reasons combined. You have to take some type of financial risk. One way is the OPM method: use Other People's Money. Unfortunately, it takes a smooth talker to get other people to part with their money, so you may have to run up your credit card, or go deep into your savings. There are many ways to get the money, but it will cost money to get your idea to market.

Marketing

You *have* to sell your product. No one will buy a product that isn't sold. That takes marketing. 'Nuff said.

Ok, maybe not, I used to think this was pretty obvious, but when I started a business helping people get stuff to market, I found this is often the most ignored part of getting a business off the ground.

If you build a better mouse trap, the world will not pound a path to your door without an infomercial, a store or a way to know it exists. You will need to be a salesman of some type to get your idea off the ground. If someone doesn't buy it, you don't have a product, just another idea that didn't go anywhere. The patent libraries are full of mouse traps that you can't buy anywhere.

Conclusion

So, will you be the next Bill Gates? Just think of a product that everyone wants. Get a couple of rich relatives to put in a good word and pitch in a few bucks, and who knows. If your timing is right it just might happen. If not, I still have that book for sale.

Thumb Rules

- Have a good idea (this is the easy part).
- Consider the market potential.
- Needs to work well.
- Looking good helps.
- Timing is everything.
- It won't happen without funds.
- You need to sell it.

The Last Word

So there it is, everything I ever learned as an engineer and a manager all distilled into a few dozen thumb rules that I hope will help you in your journey through the world of "sparkies." It is likely that somewhere someone will say, "But what about _____ rule?" I use that all the time. Well, what are second editions for? If it goes well with this one, maybe there will be more.

I hoped you liked it, I hope it helps and, most of all, I hope these thumb rules bring you success.

Darren C. Ashby

Appendix

One of my favorite areas in a physics book is the inside cover. It is where all the good stuff is distilled into the fundamentals. I couldn't call this book complete without creating a similar section.

Code Words

Here are some terms that you may or may not know—words that are often used in the realm of electronics, but will typically cause a look of confusion on any nonengineer who accidentally overhears a conversation between a couple of sparkies. They are a secret code, usually short to be more efficient, and sometimes intended to baffle the boss or at least make him wonder what you are really talking about. They have been selected at will based on looking at my own secret decoder ring and deciding what was OK to reveal without risking lynching by my fellow engineers.

AC – Alternating current, or current flow that switches back and forth. This is the type of power that comes in on the line to your house and is available at a common outlet.

Back EMF – EMF means *electromotive force*. This is used to describe the voltage generated when you spin the armature of a DCPM motor. It is also used to describe the voltage generated at the connections of an inductor when you stop pushing current through it and the magnetic field collapses. Since they are both voltages caused by a changing magnetic field, it makes some sense.

Bias – This is a widely used term in electronics. Bias can refer to the voltage applied to a circuit. For example, a DC bias or offset is a way of shifting an AC signal from one level to another, such as biasing a circuit or component to a level where you get a predictable behavior. You can bias the input of a transistor, for example.

BS – Come on, everyone knows what "BS" means!

Bulk Cap – This means a large value capacitor, usually 1 µf or bigger, commonly 100 µf to 0.1 f. Usually an electrolytic cap, not typically good at fast frequencies, has plenty of current capability.

Cap – Capacitor, a plate-like unit with a space of something that won't conduct electric between the plates. A cap has capacity to store energy in the form of an electric field.

Chip – Slang for IC; you will often hear engineers refer to ICs as chips.

Current – Describes the movement of electrons, commonly thought of as a flow. In the water analogy this is the amount of water moving. Amp is the basic unit of current in an electrical circuit. Common symbols are *I* and less often *A*.

DC – Direct current, or current flow that goes in only one direction. This is the type of power that comes from a battery. It is the type of power computers, and most electronics, use internally in their circuits.

DCPM – Short for *direct current permanent magnet* motor, these little guys are everywhere.

DMM – Darn Meter won't Measure, a cuss word often let loose when an engineer is yet to discover the fuse is blown in his digital multimeter. Usually precedes stalking off to the lab to find a screwdriver since you have to tear the whole meter apart just to replace a fuse.

Drain – Related to "sink," the drain is usually the connection on a device that "sinks" current into ground.

Drive – To drive a part means to apply current and voltage to make the part do what you want. You drive a load. If asked what a ____ is capable of driving, it means how much can it sink and source.

Duty Cycle – Duty cycle is a percentage of on time vs. off time—how much time the component is on duty, so to speak. If a motor has a 30% duty cycle, that means it is being used 30% of the time. The other 70% of the time it is off.

EPROM – Way back when our PROMs only had one E, you had to erase them with light. Oh yeah, it means *erasable programmable read only memory*.

EMI – *Electromagnetic interference* is anything and everything that interferes with an electric or electronic circuit. It is sometimes attributed to supernatural causes by superstitious engineers.

EULA – Everyone is Unable to take Legal Action if this product destroys your data. If you have never agreed to a EULA and you own this book…well…wow. I am left at a complete loss trying to come up with a quirky remark.

FAE – Field application engineer.

Flux – Flux or resin is an acid either applied separately or in the core of the solder. It cleans the joint when heated to help the solder stick better.

Forward Bias – Refers to the biasing of a diode; when forward biased, a diode passes current.

Freewheel Diode – This is a reverse biased diode hooked up in parallel with a motor. It is there to capture the inductive current that is generated as the magnetic field collapses.

Gnd, Vss – The voltage reference point. Usually you connect one lead of a measuring instrument to this point. It is also the place all the current returns to (conventional flow again) that comes from Vcc. In electron flow terms, it is the point that spews forth electrons.

Ground – While often used interchangeably with circuit gnd, ground should be thought of differently. Ground is the dirt under your feet into which you drive a big stake and hook it up to the exposed metal (and sometimes the gnd) of your circuit. This is done for safety reasons.

HW – Abbreviation for hardware.

IC – Integrated circuit. This is a device that made up of a combination of diodes and transistors and other basic parts etched into a silicon base, used to make things as simple as switches and as complex as the Intel Pentium 4 in your PC.

Impedance – Seen as a Z in many equations. Think of this as resistance that takes frequency into account. Used in conjunction with inductors and capacitors.

Inductor – A coil of wire at its most fundamental; it can store energy in the form of a magnetic field. When a magnetic field changes, it induces current to flow in a wire. The coils concentrate the magnetic field.

Iron – Soldering iron used to create solder junctions.

ISA – *Intuitive signal analysis*, the first acronym of my own invention. I figure if I ever want to be a famous engineering writer, I'd better have one or two acronyms to my name.

Junction – The place two semiconductors come together.

Ladder Logic – A type of programming method or language, its name comes from the ladder-like appearance of the diagram used to describe the program.

Lead – A pin on an electronic part such as an IC that is used to connect the part to the PCB.

Leaky Cap – A nonperfect capacitor that allows some amount of DC current to pass.

Linear – A term often used in conjunction with supply or control. A linear control is one that controls voltage to a part continuously. The part controlling this is going to dissipate energy based on the voltage across it and the current through it. It is typically an inefficient way to drive a load, as the power that is not used is turned into heat.

Load – Something that takes power, needing both current and voltage, to drive. A resistor that returns current from Vcc to Gnd is a load.

MAMA – Management Always chasing the Market Around. My own personal acronym, if you want to be successful in the world of engineering, you have to invent an acronym or two. Chalk another one up for me!

MCU – Microcontroller, which is like a CPU, but less powerful with more stuff built in.

NO, NC – Pronounced nnnn ohhh and nnn seee, this is a cryptic abbreviation for the typical state of a switch or relay connection. See, even in engineering, NO doesn't always mean no.

OS – Operating system.

OTP – *One time programmable*. Before flash became the memory of choice in embedded micros, one chance was all you got. There are still a few out there, but you are probably in some really high volumes if you are using these, and it's likely you are into masked parts as well.

Pad – The point on a PCB of bare copper where the leads of a part are connected by solder to a trace.

PCB or PWB – Printed circuit board or printed wiring board, this is a composite material usually stiff like a board on which a circuit is laid out, creating connections between components.

PLD – Programmable logic device. Take a whole bunch of memory cells, a slew of logic gates, a bunch of multiplexers and a way to configure it all. Cram it into a single IC. At the end of all this, you get a product that can do a whole bunch of state machine and logic stuff. You can even make MCUs out of them as seen in sister products like the FPGA.

PM – Permanent magnet.

Pointy Hair – We have Scott Adams to thank for this unique reference that we can now use to refer to our boss.

Power – The combination of voltage and current is power. This is what turns the lights in on in your house. The unit for power is the *watt*. The common symbol is W. Watts can be converted to horsepower; it takes 746 watts to make one horsepower. Another symbol you may see that is loosely related to watts is VA or volt amps. This is a symbol usually used in power supply systems referring to AC power; it is equivalent to watts only when the current and voltage match phases.

Power Component – The term "power component" is commonly used to refer to parts that handle a large amount of current or high voltage. Of course the words large and high are relative. It means a current large enough you need to worry about things like heat, and voltage high enough that it will do more than tickle a little if you touch it.

Power Device – Term commonly used to refer to semiconductor devices such as FET and transistors that take a small low-power input signal and amplify it into a high-power signal. Power devices usually need to be meticulously handled in your design to avoid overheating. They often have a surface that is designed to be coupled to a heat sink to manage the power dissipated as they operate.

Pull-Up – This is a resistor from an input line to Vcc. In the absence of any other current flow, it "pulls" the voltage at that node to Vcc.

Pull-Down – This is a resistor from an input line to gnd. In the absence of any other current flow, it "pulls" the voltage at that node to Gnd.

PWM – Pulse width modulation, a digital method of controlling a voltage level. The percentage of time on versus time off determines the amount of power applied to the load.

R – Pronounced arrrrr, as what is the arrr of that puppy, means resitance, something that resists the flow of current proportionally to the voltage; it is the "R" in Ohm's Law.

Rail – The voltage limit to which an output can swing. The top rail is the highest positive voltage it can get to, and the bottom rail is the lowest voltage it can get to. This is not necessarily the same as the power supply. Some devices cannot get the output to reach Vcc or Gnd in the circuit. When the output is at these limits, it is common to say it is railed.

Rectify – Rectify or rectification is the process of turning AC power into DC power.

Reverse Bias – This a specific case of biasing, usually referring to a diode. When a diode (or diode type junction in a component) is reverse biased, the diode blocks current flow.

Sink – No, not the kitchen sink, but it does act a little like a drain. This is usually used in a phrase like "How much can that sink?" It means how much current is capable of going into ground through that part.

Solder – A material used to make electrical connections. It is heated to create that connection.

Source – This is usually used in a phrase like "How much can that source?" It means how much current is capable of coming out of that part. Both sink and source assume conventional current flow terminology from positive to negative.

Sparky – Widely used slang term to refer to an electrical engineer, at least in the world of Darren. (We tried to assign the term "wrench" to the MEs but it just doesn't have the same ring to it.)

State Machine – This is a computing device that looks at the state of the inputs to determine the output. More complex forms of this device feedback outputs to the input and/or maintain memory of certain inputs.

SW – Abbreviation for software.

Switcher – A cousin to the linear control or supply above is the switcher. The switching control is digital in nature. Somewhere in the system is a switch that turns on and off cycling power to the load. The amount of time on vs. off is called the duty cycle. It is defined as a percentage. Often there is an inductive or capacitive component in or attached to the load that filters the frequency of the switching device to smooth out the voltage or current to the load.

Switch Mode – Digital control of a device like a transistor or FET, for example. The part is turned either all the way on or off, like a switch—hence, switch mode control. Using a device like this in applications like a switching power supply help make it more efficient, as less heat is created when a part is not in the linear region of operation.

Threshold – In electronics, the threshold is a voltage level that, when crossed, changes the output state of a logic circuit from 1 to 0 or vice versa.

Tinning – This refers to applying solder to the tip of an iron or to a wire to help heat transfer.

Trace – These are the little green lines you see on a PCB. They are made of copper and are the wires that connect the parts together. Trace can also refer to a method of troubleshooting software.

Vcc, Vdd – This is the voltage source in the circuit. In conventional flow terms, it is the place all the positive holes come from. In electron flow terms, it is the place where all the electrons try to get to.

Via – A via is a hole in a PCB that on some PCBs is coated with copper. It is used for two reasons, either to create a connection between a top trace and a bottom trace and or to create a hole in which a part lead can be inserted and be soldered to the PCB.

Voltage – This is the potential of the available electrons. Using the water analogy this is the pressure the current is under to move. The unit for voltage is the *volt*. Common symbols are *V* and *E*.

References

Everyone learns from many sources, while I can say my take on these engineering basics is as original as I am aware of, I must say that I have read many books that helped to gain that insight. In general they will delve deeper into the topic than I did, so you may find them useful. Here is a list of those books, with maybe a few comments to help you to decide if you want to read them yourself.

Yellow Control Theory, Fundamentals of Automatic Control, by Robert C. Weyrick, McGraw Hill, ISBN 0-07-069493-1. Good read, helped me understand control theory.

DC Motors Speed Controls Servo Systems – I like to call it the pink motor book due to an interesting choice of color for the cover, and I highly recommend it for anyone who is working with DC motors. It's heavy on the equations, but a good source for understanding all the complexities of motors.

Pink Motor Book, DC Motors Speed Controls Servo Systems, The Electro-Craft Engineering Handbook, by Reliance Motion Control, Inc.

Grounding and Shielding Electronic Systems, by Dr. Tom Van Doren, University Missouri Rolla, Van Doren Company, Rt 6 Box 319, Rolla, Mo 65401, Ph 314-341-4097.

Intuitive IC Op Amps, by Tom Frederickson. This classic paperback book was originally published in 1984. The book describes how op-amps work and how they are used, from a practical, commonsense perspective. It is currently out of print. However, you may be able to find it in university libraries or by browsing the Internet. As of March 2005, the book was also available from Rector Press.

This book was written by the inventor of the most widely used op-amp in the world, the LM324. This book gave me the first hint that op-amps should be easy to use, not hard.

Useful Websites and Magazines

Following are some websites you may find useful.

ePanorama.net is a website dedicated to offering information on electronics found on the web – www.epanorama.net

"Circuit Cellar" Magazine – www.circuitcellar.com

Resistor Color Codes

I can never remember what all the color codes are, I get the basic 10, 100, 1K, etc. pretty good, and I can usually identify the 4.7 stuff, but all the rest are beyond me. So here is a quick reference guide if you need to look them up.

Please reference the accompanying CD-ROM for more information.

Capacitor Values

The easiest ones to read are the electrolytic type, because it is printed right on the side. Here is a reference for the other markings.

Please reference the accompanying CD-ROM for more information.

Equations

A quick reference of all the equations we have covered in this book can be found on the accompanying CD-ROM for more information

Thumb Rules Revisited

A quick reference of the thumb rules for all the topics covered can be found on the accompanying CD-ROM for more information.

About the Author

Darren Coy Ashby is a self-described "techno geek with pointy hair." He considers himself a Jack of all trades, master of none. He figures his common sense came from his dad and his book sense from his mother. Raised on a farm and graduating from Utah State University seemingly ages ago, he has 16 years of experience in the real world as a technician, engineer and a manager. He has worked in diverse areas of compliance, production, testing and his personal favorite, R&D.

He jumped at a chance some years back to teach a couple of semesters at his alma mater. For about two years, he wrote regularly for the online magazine "chipcenter.com." He is currently the head of electronics R&D at the world's largest fitness company. His passions are snowmobiles, boats, motorcycles and pretty much anything with a motor in it. When not at his day job, he spends most his time with his family and a start-up R&D consulting firm he initiated a coupled of years ago.

He lives with his beautiful wife, four strapping boys and cute little daughter next to the mountains in Richmond, Utah. You can email him with comments, complaints and general ruminations at dashby@raddd.com.

Index

F

M

P

Q

R

T

W

TERM

This Agreement will remain in effect until terminated pursuant to the terms of this Agreement. You may terminate this Agreement at any time by removing from Your system and destroying the CD-ROM Product. Unauthorized copying of the CD-ROM Product, including without limitation, the Proprietary Material and documentation, or otherwise failing to comply with the terms and conditions of this Agreement shall result in automatic termination of this license and will make available to Elsevier Science legal remedies. Upon termination of this Agreement, the license granted herein will terminate and You must immediately destroy the CD-ROM Product and accompanying documentation. All provisions relating to proprietary rights shall survive termination of this Agreement.

LIMITED WARRANTY AND LIMITATION OF LIABILITY

NEITHER ELSEVIER SCIENCE NOR ITS LICENSORS REPRESENT OR WARRANT THAT THE INFORMATION CONTAINED IN THE PROPRIETARY MATERIALS IS COMPLETE OR FREE FROM ERROR, AND NEITHER ASSUMES, AND BOTH EXPRESSLY DISCLAIM, ANY LIABILITY TO ANY PERSON FOR ANY LOSS OR DAMAGE CAUSED BY ERRORS OR OMISSIONS IN THE PROPRIETARY MATERIAL, WHETHER SUCH ERRORS OR OMISSIONS RESULT FROM NEGLIGENCE, ACCIDENT, OR ANY OTHER CAUSE. IN ADDITION, NEITHER ELSEVIER SCIENCE NOR ITS LICENSORS MAKE ANY REPRESENTATIONS OR WARRANTIES, EITHER EXPRESS OR IMPLIED, REGARDING THE PERFORMANCE OF YOUR NETWORK OR COMPUTER SYSTEM WHEN USED IN CONJUNCTION WITH THE CD-ROM PRODUCT.

If this CD-ROM Product is defective, Elsevier Science will replace it at no charge if the defective CD-ROM Product is returned to Elsevier Science within sixty (60) days (or the greatest period allowable by applicable law) from the date of shipment.

Elsevier Science warrants that the software embodied in this CD-ROM Product will perform in substantial compliance with the documentation supplied in this CD-ROM Product. If You report significant defect in performance in writing to Elsevier Science, and Elsevier Science is not able to correct same within sixty (60) days after its receipt of Your notification, You may return this CD-ROM Product, including all copies and documentation, to Elsevier Science and Elsevier Science will refund Your money.

YOU UNDERSTAND THAT, EXCEPT FOR THE 60-DAY LIMITED WARRANTY RECITED ABOVE, ELSEVIER SCIENCE, ITS AFFILIATES, LICENSORS, SUPPLIERS AND AGENTS, MAKE NO WARRANTIES, EXPRESSED OR IMPLIED, WITH RESPECT TO THE CD-ROM PRODUCT, INCLUDING, WITHOUT LIMITATION THE PROPRIETARY MATERIAL, AN SPECIFICALLY DISCLAIM ANY WARRANTY OF MERCHANTABILITY OR FITNESS FOR A PARTICULAR PURPOSE.

If the information provided on this CD-ROM contains medical or health sciences information, it is intended for professional use within the medical field. Information about medical treatment or drug dosages is intended strictly for professional use, and because of rapid advances in the medical sciences, independent verification f diagnosis and drug dosages should be made.

IN NO EVENT WILL ELSEVIER SCIENCE, ITS AFFILIATES, LICENSORS, SUPPLIERS OR AGENTS, BE LIABLE TO YOU FOR ANY DAMAGES, INCLUDING, WITHOUT LIMITATION, ANY LOST PROFITS, LOST SAVINGS OR OTHER INCIDENTAL OR CONSEQUENTIAL DAMAGES, ARISING OUT OF YOUR USE OR INABILITY TO USE THE CD-ROM PRODUCT REGARDLESS OF WHETHER SUCH DAMAGES ARE FORESEEABLE OR WHETHER SUCH DAMAGES ARE DEEMED TO RESULT FROM THE FAILURE OR INADEQUACY OF ANY EXCLUSIVE OR OTHER REMEDY.

U.S. GOVERNMENT RESTRICTED RIGHTS

The CD-ROM Product and documentation are provided with restricted rights. Use, duplication or disclosure by the U.S. Government is subject to restrictions as set forth in subparagraphs (a) through (d) of the Commercial Computer Restricted Rights clause at FAR 52.22719 or in subparagraph (c)(1)(ii) of the Rights in Technical Data and Computer Software clause at DFARS 252.2277013, or at 252.2117015, as applicable. Contractor/Manufacturer is Elsevier Science Inc., 655 Avenue of the Americas, New York, NY 10010-5107 USA.

GOVERNING LAW

This Agreement shall be governed by the laws of the State of New York, USA. In any dispute arising out of this Agreement, you and Elsevier Science each consent to the exclusive personal jurisdiction and venue in the state and federal courts within New York County, New York, USA.